Understanding Complex Systems

Springer Complexity

Springer Complexity is an interdisciplinary program publishing the best research and academic-level teaching on both fundamental and applied aspects of complex systems—cutting across all traditional disciplines of the natural and life sciences, engineering, economics, medicine, neuroscience, social and computer science.

Complex Systems are systems that comprise many interacting parts with the ability to generate a new quality of macroscopic collective behavior the manifestations of which are the spontaneous formation of distinctive temporal, spatial or functional structures. Models of such systems can be successfully mapped onto quite diverse "real-life" situations like the climate, the coherent emission of light from lasers, chemical reaction-diffusion systems, biological cellular networks, the dynamics of stock markets and of the internet, earthquake statistics and prediction, freeway traffic, the human brain, or the formation of opinions in social systems, to name just some of the popular applications.

Although their scope and methodologies overlap somewhat, one can distinguish the following main concepts and tools: self-organization, nonlinear dynamics, synergetics, turbulence, dynamical systems, catastrophes, instabilities, stochastic processes, chaos, graphs and networks, cellular automata, adaptive systems, genetic algorithms and computational intelligence.

The three major book publication platforms of the Springer Complexity program are the monograph series "Understanding Complex Systems" focusing on the various applications of complexity, the "Springer Series in Synergetics", which is devoted to the quantitative theoretical and methodological foundations, and the "Springer Briefs in Complexity" which are concise and topical working reports, case studies, surveys, essays and lecture notes of relevance to the field. In addition to the books in these two core series, the program also incorporates individual titles ranging from textbooks to major reference works.

Indexed by SCOPUS, INSPEC, zbMATH, SCImago.

Hassan Qudrat-Ullah
Editor

Understanding the Dynamics of New Normal for Supply Chains

Post COVID Opportunities and Challenges

Springer

Editor
Hassan Qudrat-Ullah
School of Administrative Studies
York University
Toronto, ON, Canada

ISSN 1860-0832 ISSN 1860-0840 (electronic)
Understanding Complex Systems
ISBN 978-3-031-07332-8 ISBN 978-3-031-07333-5 (eBook)
https://doi.org/10.1007/978-3-031-07333-5

This Springer imprint is published by the registered company Springer Nature Switzerland AG
The registered company address is: Gewerbestrasse 11, 6330 Cham, Switzerland

This book is dedicated to Usman Hassan Ali, my grandson, with the hope and prayers that he would become a scholar of Islamic teachings to overcome the challenges of today's world.

Preface and Acknowledgements

Decision-making in supply chains in technology-intensive and ever-in-turbulent business environments is a complex task at best. Comes COVID-19 simply adds to this complexity. Given the global trend of focusing on technology-based solutions, supply chain design and management has become a dynamic problem. Understanding the dynamics of supply chains due to multiple stakeholders' perspectives, rising environmental concerns about CO_2 emissions and unpredictable external shocks like COVID-19 become pronounced. Therefore, the use of dynamic modeling and strong theoretical frameworks-based analysis for supply chains design and management has seen phenomenal growth during the past several decades.

The primary aim of this book is to disseminate the roles and applications of various modeling approaches enriched with strong theoretical frameworks aimed at improving the usefulness of dynamic modeling-based solutions in supply chains design and management. Decision. The key focus is on the applications of system dynamics, econometric, multi-criteria models, and theory-based decision-making frameworks in service of supply chains design and management in the highly uncertain and fast-paced business environments of the twenty-first century. Invitations for contributions were sent all around the globe. Several well-known and prominent scholars were also specially invited to contribute. Each prospective contributor was initially asked to prepare a two to three-page long abstract of their chapter. These proposals were reviewed by the editor. Suggestions were made to prepare the full papers. The submitted papers were then reviewed by independent reviewer panels. Each panel consisted of three members—the editor and two independent experts in the field. The final acceptance/rejection decisions were made by the editor based on the revised papers submitted by the contributors.

The book contains three parts. Part I, "Introduction to Supply Chain Dynamics and Post-COVID Implications" has one chapter. It introduces key aspects of major modeling approaches and presents an overview of all nine chapters of this book. Part II of the book, "Understanding Supply Chain Dynamics -Theoretical Perspectives," consists of three chapters that deal with unique theoretical perspectives for enhancing the management and performance of post-COVID supply chains including *Post-COVID Supply Chain Dimensions Proposed Shift: A Qualitative Dynamic*

Model, Covid-19 Cases, and Supply Chain Disruptions, and Characteristics and Capabilities of a Successful Supplier: A Conceptual Model.

Next, Part III, "The Innovative Modeling-Based Solutions for Industry-Specific Supply Chains," showcases the state-of-the-art molding-based six innovative solutions for industry-specific (e.g., healthcare, transport, pharmaceutical, livestock, and automotive) supply chains including Sustainable Outcomes Through The Structured Forward Supply Chain: A System Dynamic Approach. Using System Dynamics modeling to frame environmental voluntary commitment programs in the transport industry in the post-COVID situation: the French experience, Waste Processing Scenarios to Support Sustainable Environmental Development Using System Dynamics, Multi-criteria analysis of disruption risks for supply chains due to pandemics, Botswana state-owned enterprises (SOE): A review of lessons learned from supply chain management best practices during COVID-19, and Collaboration Model between Buyer and Supplier: An Empirical Assessment of Indonesian Pharmaceutical Industry. Finally, the last Part IV lists the thematic bibliography to advance future research about post-COVID supply chains. The unique feature of this book is the explicit inclusion of the section about, "*Implications for Post-COVID Supply Chains Practice*".

We are grateful to the authors of the various chapters for their contributions. It had been a bit long process from the initial outlines to developing the full chapters and then revising them in the light of reviewers' comments. We sincerely acknowledge the authors' willingness to go through this process. We also acknowledge the work and knowledge of the members of our review panels, many of which had to be done at short notice.

Thanks to all the people at Springer, USA especially Niko Hisako with whom we corresponded for their advice and facilitation in the production of this book.

Jayanthi Krishnamoorthi, Scientific Publishing Services Private. Prepared a camera-ready copy of the manuscript with her usual professionalism and cooperation and we wish to record our thanks to her.

Finally, I am grateful to my family, Tahira (for her unwavering, incredible, and selfless support all the times), Anam (for her professional proofreading and spiritual perspective on things around us), Ali (for not sparing time from his kingdom), Umer (for his occasional smiles on my work and his work too), Umael (for bearing with me on being away and asking about his recitation of Quran practice), and my mother, Fazeelat Begum (for her sacrifice, support, and prayers all along despite all the odds)—all the very source of my inspiration and desire to embark on this journey. It would be unfair not to acknowledge the constant and consistent prayers and cares she extends to me, Saira Bano, my mother-in-law.

Toronto, Canada
March 2022

Hassan Qudrat-Ullah

Contents

Contributors

Bilek Gisele Mendy Business Management, IAE Pau Bayonne, LIREM, Université de Pau Et Des Pays de L'Adour, Pau, France

Bivona Enzo Business Management, University of Palermo, Palermo, Italy

Hendrawan Rully Agus Department of Information Systems, Institut Teknologi Sepuluh Nopember (ITS), Surabaya, Indonesia

Neto J. Martino Department of Production, Sao Paulo State University, Guaratingueta, SP, Brazil

Putri Dinda Meidianti Kusuma Department of Information Systems, Institut Teknologi Sepuluh Nopember (ITS), Surabaya, Indonesia

Qudrat-Ullah Hassan School of Administrative Studies, York University, Toronto, ON, Canada

Rahmawati Ulfa Emi Department of Information Systems, Institut Teknologi Sepuluh Nopember (ITS), Surabaya, Indonesia

Salomon Valerio Antonio Pamplona Department of Production, Sao Paulo State University, Guaratingueta, SP, Brazil

Sebata Dennis Monarch Business School Switzerland, Zug, Switzerland

Shamsuddoha Mohammad Supply Chain Management, Western Illinois University, Macomb, IL, USA

Suryani Erma Department of Information Systems, Institut Teknologi Sepuluh Nopember (ITS), Surabaya, Indonesia

Szulanski Fabián Universidad del CEMA, Buenos Aires, Argentina

Szulanski Fabián UCEMA University, Buenos Aires, Argentina

York Oxford Monarch Business School Switzerland, Zug, Switzerland

Yunus Erlinda Nusron Operations Excellence Department, Sekolah Tinggi Manajemen PPM, Jakarta, Indonesia

Part I
Understanding Supply Chain Dynamics: An Introduction

Chapter 1
Introduction to Supply Chain Dynamics and Post-COVID Implications

Hassan Qudrat-Ullah

Abstract Understanding the dynamics of supply chains in highly uncertain and dynamic business environments is an essential prerequisite for effective and efficient managerial decision making in most business organizations. This chapter presents the case for this book and provides an overview of its content.

Keywords Dynamics of supply chains · Supply chain risks · Sustained growth · COVID-19 · Sustainability practices · Supply chain operations · Supply chain resilience · Modeling approaches · System Dynamics · Econometric models

1.1 Introduction

Faced with the challenges of operating under a highly uncertain, unpredictable, and disruptive business environment (e.g., due to COVID-19 like events) appears to be the new normal for supply chains. As a result, how to minimize the overall supply chain risks and sustained growth, is the key question that today's businesses have to address, subject to the available but limited resources of the businesses. Multiple, conflicting, and dynamic objectives of stakeholders including suppliers, manufacturers and service providers, and retailers add to the complex nature of decisions that modern-day managers of supply chains face.

In the face of COVID-19, organizations are in search of models and strategies that can help to overcome the ensuing challenges and risks of their supply chains. Collaborative approaches with suppliers, sustainability practices in supply chain operations, and building supply chain resilience can play a critical and supportive role in this era (Kiers et al., 2022; Paul et al., 2021; Sajjad, 2021). Businesses with diverse, competent, talented, and engaged human capital across the entire network are likely to build and sustain their supply chain resilience capability. Increased supply chain resilience and responsiveness can help organizations cope with disruptive and unpredictable demand and supply imbalances (Barbieri et al., 2020). Therefore, here

H. Qudrat-Ullah (✉)
School of Administrative Studies, York University, Toronto, ON, Canada
e-mail: hassanq@yorku.ca

H. Qudrat-Ullah (ed.), *Understanding the Dynamics of New Normal for Supply Chains*, Understanding Complex Systems,
https://doi.org/10.1007/978-3-031-07333-5_1

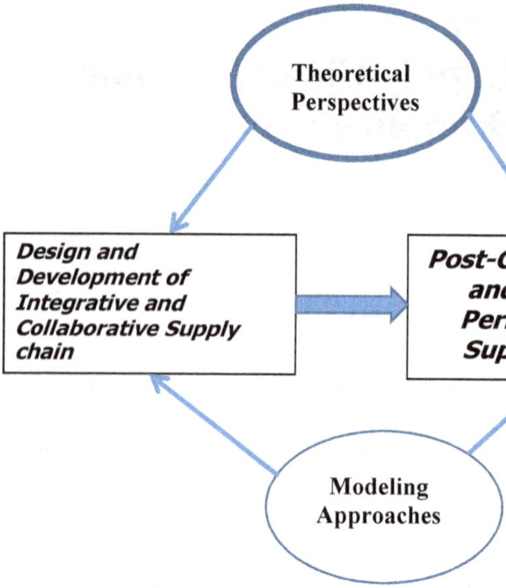

Fig. 1.1 Perspectives and modeling approaches for post-COVID supply chains

are two topical issues about the development and sustained growth of supply chains across the domains in the Post-COVID era (as shown in Fig. 1.1) that are addressed in this book:

(i) How do theoretical perspectives (e.g., emerging factors for suppliers' selection, shifts in how strategy [to cope with high-stake disruptions] is designed, and reasoning behind varying experiences during supply chain disruptions) provide actionable insights and decisions about the design, development, and growth of integrative and collaborative supply chains in a Post-COVID era?
(ii) How do various modeling approaches (e.g., system dynamics, econometric, and case-based analysis) provide innovative solutions to the issues about the Post-COVID design, development, and sustained operations of supply chains? What are the key implications for the practice of supply chains in this era?

Consequently, in the post-COVID era, where abrupt and sudden variations in the demand–supply supply of products and services appear to become a norm, supply chain issues and problems require a systematic and integrated approach. Modeling and simulation in general and system dynamics modeling, in particular, have the potential and capabilities to cope with the complexity of supply chain-related problems in these uncertain times. Therefore, the primary objective of this book is to present the latest decision making perspectives, tools, techniques, and insightful and innovative modeling solutions that decision makers can utilize to overcome the challenges and risks of their supply chains.

To provide some unique theoretical perspectives and innovative modeling-based solutions for Understanding the Dynamics of the New Normal of Supply Chains, we issued the call for contributions in this volume. Specifically, we sought help from the system dynamics modeling community. Consequently, several different examples of perspectives and modeling approaches, with a common unifying goal of "improving the performance of post-COVID supply chains" are provided in this volume.

1.2 Methodology

In our call for contributions on "Understanding the Dynamics of New Normal of Supply Chains," we went through various email lists of professional bodies. We also posted the call for chapters on message boards of a few international conferences on the related topics. Personal invitations were sent to target authors as well. We received seventeen "Two-to-Three page" abstracts as an expression of interest. Based on the initial screening by our review panel, the authors of eleven chapters were invited to submit the complete chapter. All eleven chapters received from the contributors went through a double-blind review process. The reports from the independent reviewers were sent to the authors to address the issues and incorporate the suggestions made by the reviewers. Only nine chapters made it to the final stage of acceptance. The final versions of these nine chapters have been edited and included in this volume.

1.3 Research Categories

We applied a thematic approach to classifying our contributions to this book. The chapters thus complied are classified into four categories following the structure of the book. The first category, the current one, presents the introduction and preview of "Understanding the Dynamics of the New Normal of Supply Chains". The second category examines the two unique theoretical perspectives for enhancing the performance of post-COVID supply chains including *Post-Covid Supply Chain Dimensions Proposed Shift: A Qualitative Dynamic Model, Covid-19 Cases, and Supply Chain Disruptions, and Characteristics and Capabilities of a Successful Supplier: A Conceptual Model.*

Next, the third category showcases the state-of-the-art molding-based six innovative solutions for industry-specific (e.g., healthcare, transport, pharmaceutical, and automotive) supply chains including *Sustainable Outcomes Through The Structured Forward Supply Chain: A System Dynamic Approach. Using System Dynamics modeling to frame environmental voluntary commitment programs in the transport industry in the post-COVID situation: the French experience, Waste Processing Scenarios to Support Sustainable Environmental Development Using System Dynamics, Multi-criteria analysis of disruption risks for supply chains due to pandemics, Botswana state-owned enterprises (SOE): A review of lessons learned from supply chain management best practices during COVID-19, and Collaboration Model between Buyer and Supplier: An Empirical Assessment of Indonesian*

Pharmaceutical Industry. Finally, the last category lists the thematic bibliography to advance future research about post-COVID supply chains.

1.4 The Unique Theoretical Perspectives for the Post-COVID Supply Chains

Here we present two unique perspectives for enhancing the performance of post-COVID supply chains. These perspectives shed light on important actionable strategies and decisions about sustainable supply chains. For instance, what 'shifts" in supply chains strategies are needed, why "supplier selection" is so critical, and how some businesses and "cases" did well during this disruptive era.

1.4.1 Shifts in Supply Chain Strategies

When it comes to the unique perspectives for sustainable supply chains, the R&D technologists' perspective presents some interesting insights. In Chap. 2, "**Post-Covid Supply Chain Dimensions Proposed Shift: A Qualitative Dynamic Model**," Fabián Szulanski, reviews and suggests "shifts" in post-Covid supply chain strategic experimentation. Considering the Covid-19 pandemic as a Black Swan, the author argues that the impact of this can change most of the methods and manners of how value is created. Similarly, this disruptive impact affects how firms collaborate, strategize, and make operational and strategic decisions. Citing some examples of supply chain decision making during the lockdown, the author argues that there are shifts expected in how supply chain firms collaborate, strategize, and make strategic decisions. Post-COVID supply chain performance-related implications are discussed by the author. Finally, some important future research avenues are presented especially experimental and empirical research is called.

1.4.2 Suppliers Selection in Post-COVID Era

The traditional approach to supplier selection where "cost-competitive" criteria alone were at work, can hardly help the organization cope with the unprecedented damages and impacts of COVID-19 like a pandemic or a similar external shock event. Hassan Qudrat-Ullah, in Chap. 4, "**Characteristics and Capabilities of a Successful Supplier: A Conceptual Model**," of this book, present an integrative model that links both the traditional and emerging factors. He posits that suppliers play a fundamental role in the success of any supply chain. "The damaging and disruptive effects of pandemics (i.e., COVID-19) on businesses and their supply chains have made the role of suppliers even more critical and pronounced," he asserts. In this chapter, based on a critical review of the literature about suppliers section, the author develops a conceptual model that links both traditional and emerging criteria for suppliers' selection. Then, the developed model is applied to the case of the automotive industry. His analysis shows that in a Post-COVID-19 era, suppliers need to invest and develop competencies, capabilities, and resources in human capital, communication skills, information technologies, and strategic commitment. He also provides insights and implications for managing the supply chain risks in a Post-COVID era.

1.5 The Innovative Modeling-Based Solutions for Industry-Specific Supply Chains

After the reader of this book has learned about two unique perspectives unified with a common goal: improving the performance of post-COVID supply chains, here in this section we present six state-of-the-art solutions, rich with actionable insights and implications, for industry-specific supply chains including the (i) wind power industry, (ii) essential consumer goods industry, (ii) food industry, (iv) tourism industry, and (v) timber industry. Insights and lessons learned from these industry-specific supply chains, however, apply equally well to other industries and jurisdictions.

1.5.1 Sustainable Outcomes and the Structured Forward Supply Chains

This chapter deals with the sustainability issues for poultry livestock using Bangladesh as a case study. Utilizing the system dynamics modeling approach, the author, Mohammad Shamsuddoha, in this Chap. 5, "**Sustainable Outcomes Through The Structured Forward Supply Chain: A System Dynamic Approach**" integrates and analyzes sustainability and supply chains. According to the author, for modern businesses, in the face of highly competent business environments and increasing need for economic, social, and environmental sustainability, the concepts of sustainable supply chains have become critical. However, not all the sectors of economy are receiving the same attention to this aspect of businesses. For instance, the author of this chapter claims that literature is scant about sustainable supply chains for the poultry livestock sub-sector of the economy. Instead, the poultry livestock industry is faced with non-standard and unstructured supply chain processes, minimal attention is paid to the concept of sustainability, and often poultry wastes are rarely recycled. Like other industries, the poultry livestock industry and its supply chain are severely impacted by the COVID-19 pandemic. Utilizing system dynamics and case study approaches, this quantitative study used the positivist paradigm and design science methodology to model and analyze the performance of several real-world supply chaincases. Post-COVID implications for the management of the sustainable supply chains for the poultry livestock industry are discussed in this chapter.

The findings of this chapter revealed that supply chain integration could provide economic and social sustainability and a structured manufacturing process to support the research objectives and questions. At the end of the chapter, he discusses briefly the pandemic effects on the supply chains of the poultry industry to determine the future research directions.

1.5.2 Environmental Voluntary Commitments and Transport Sector Supply Chains

Transport sector supply chains across the globe are concerned with their ongoing carbon footprint. Continuing with the use and application of system dynamics methodology, Bivona and Bilek, in Chap. 6, "**Using System Dynamics modeling to frame environmental voluntary commitment programs in the transport industry in the post-COVID situation: the French experience**," present development and application of a stock-and-flow model. Authors, in this interesting piece of research, advance their assertions as: (i) The success of the Engagements Volontaires pour l'Environnement (EVE) program recently launched in France in improving efficiency of transport operators and reduction of related CO_2 emissions is dependent upon how well the COVID induced crises are managed, (ii) a system dynamics model that can effectively capture the complexity the multiple relationships between public and private actors involved in the EVE program is presented as model solution, (iii) understanding the casual path among the variables, key feedback loops, and fundamental stock-and-flow structures of the EVE program, essentially a dynamic task, is pre-requite for effective decision making, and (iv) an appreciation and recognition of the understanding of the dynamic interdependences of voluntary programs such EVE that are focussed on the reduction of CO_2 emissions in the transport industry can help decisionmakers in the design, development, and implementation of successful and effective policies.

The authors of this chapter also provide some important implications for the managers including transport sector supply chains. They claim that this study can help transport operators engaged in environmental voluntary commitment programs to build a durable competitive advantage, while complying with environmental policies, particularly in the post-COVID situation.

1.5.3 *Waste Management and Disruptive Supply Chains*

Waste management has received global attention due to its increasing relevance to energy, the environment, and the economy. Erma Suryani, Rully Agus Hendrawan, Ulfa Emi Rahmawati, and Dinda Meidianti Kusuma Putri from Indonesia, demonstrate the utility of system dynamics methodology, a core claim being forwarded in this book, in creating and analyzing the waste management scenarios. Chap. 7, "**Waste Processing Scenarios to Support Sustainable Environmental Development Using System Dynamics**," provides several assertions and empirical conclusions regarding waste management and supply chains. According to the authors, waste management is a complex issue as: (i) in populous regions and cities, waste can hardly be separated from mundane activities of people resulting in unwanted waste accumulation and piles in various places, and (ii) based on the principles of sustainable environmental development, waste management paradigm can be redesigns for effective and sustainable waste management. Authors have developed a dynamic system dynamics model that fully captures the underlying internal and external factors of the waste management system systematically. They build three alternative scenarios to support the sustainable management of waste. These three scenarios are (i) recycling processed food waste, (ii) burning waste with environmentally safe practices, and (iii) improving community awareness, socialization, and training to achieve environmentally friendly treatment and management of the community. The practitioners and researchers in the supply chain of waste processing area and especially during a post-COVID period can benefit from these empirical findings and implications that are presented by the authors of this chapter.

1.5.4 *Multi-Criteria Analysis of Disruption Risks for Supply Chains*

Although the majority of empirical studies regarding the effectiveness and efficiency of supply chains in the post-COVID era are based on system dynamics modeling, we have included alternate competitive methodologies as well. This is the first study that applies a multi-criteria approach to analyze the risks and damages to supply chains due to pandemics. Authors, Jose Martino Neto and Valerio Antonio Pamplona Salomon, in Chap. 8 of this book, "**Multi-criteria analysis of disruption risks for supply chains due to pandemics**" explores ways to manage overall supply chain risks due to pandemics like COVID-10. They assert that (i) Coronavirus Disease 2019 (COVID-19) affected global economics and society, unprecedentedly, and (ii) supply chains, linking customers, manufacturers, and suppliers, are more susceptible to disruption risks when facing pandemics, like COVID-19. The authors assert that in the face of COVID-19, already several empirical and theoretical studies are conducted to design, assess and evaluate various strategies for the *supply chain management*. In an exploratory study, this chapter attempts to explore supply chain risks due to this ongoing pandemic. The key objective of this study is to develop a mathematical model *for the evaluation and assessment of pandemic-induced disruption risks in supply chains. Based on a systematic review of literature on supply chain management, authrors have developed a multi-criteria decision analysis (MCDA) model. For the assessment model, they combined two leading methods of MCDA: (i) Analytic Hierarchy Process, and (ii) Technique of Order Preference by Similarity to Ideal Solution.*

Two leading MCDA methods were combined in the development of the assessment model. This assessment moted is validated with the case study of a multinational automotive company. Based on their modeling-based analysis, they conclude (i) for efficient supply chains, the model resulted in a focus on capacity management, demand planning, and sales forecasting, to avoid risks disruptions, and (ii) for responsive supply chains, the focus shall move to operations management. They also present practical implications for the decision makers and managers of supply chains in the post-COVID period, which is characterized by increased uncertainty and risks for supply chains.

1.5.5 Managing Meat Processing, Food Service, and Poultry Supply Chains

Supply chain disruptions affect all businesses. However, small and Medium Enterprises (SMEs) and State-Owned Enterprises (SOEs) are more vulnerable than large multinational corporations. Access to resources, technologies, and talent is relatively limited for these businesses. How did they do in this COVID-19 pandemic disruption? In Chap. 8, "**Botswana state-owned enterprises (SOE): A review of lessons learned from supply chain management best practices during COVID-19**," authors, Oxford York and Dennis Sebata, present the case of SOEs in a developing country, Botswana. They advanced several important prepositions and implications including (i) that COVID-19 has created many economic disruptions in the way supply chains (SCs) are being managed, (ii) the rapid spread of the contagion has impacted severely the meat processing, food service, and poultry SCs across the globe, (iii) Botswana, in the face of COVOD-19 pandemic, adopted supply chain management best practices, and (iv) Botswana's case provide useful insights (e.g., Botswana's strong competitive advantages during crisis appear to be the quality of service, effective operations, and efficient inbound and outbound logistics) for other developing countries to better manage their supply chains during a pandemic. Although the focus of this study is on Botswana's case. The findings and insights are general enough to apply to other countries, especially to the supply chains of developing countries. Ensuring a sustainable supply of items for the food industry is the concern of all countries of the world. Policymakers can avail themselves of the findings of this study.

1.5.6 *Managing Meat Processing, Food Service, and Poultry Supply Chains*

In the final chapter of this section and this book, Chap. 10, "**Collaboration Model between Buyer and Supplier: An Empirical Assessment of Indonesian Pharmaceutical Industry**," Erlinda Nusron Yunus demonstrates the utility of the econometric modeling approach in analyzing the supply chain dynamics in the pharmaceutical industry. In the context of Indonesia, this chapter evaluates the collaboration engaged by buyers and suppliers in the pharmaceutical industry. Using 52 company data about supply chain relationships, architecture, collaboration, and performance and applying path analysis, the author finds that the supplier-buyer business relationship improves supply chain architecture, further increasing supply chain collaboration. Although this chapter applies an econometric modeling approach to better understand the dynamics of supply chains in the pharmaceutical industry, the utility of such models is limited but can be applied to analyze the complex and disruptive supply dynamics present in most of the supply chains of the post-COVID period.

1.6 Concluding Remarks

At the outset of this book project, we set the objective of this edited volume as "to present the latest decision making tools, techniques, and insightful and strong theory-based and dynamic modeling-based innovative solutions that decision makers can utilize to overcome the challenges that their supply chains face in the post-COVID ear." Both eminent and distinguished as well emerging scholars of the supply chain domain responded to our call for contributions with their, various theoretical perspectives and dynamic modeling-based unique solution-oriented research. We are successful in showcasing, here in this book, nine chapters covering a range of perspectives and modeling-based solutions unified by a common goal: to improve the performance of supply chains in the post-COVID period.

We have *two* leading contributions presenting unique perspectives including *Post-Covid Supply Chain Dimensions Proposed Shift: A Qualitative Dynamic Model and Characteristics and Capabilities of a Successful Supplier: A Conceptual Model.* These perspectives provide insights and sound reasons as to what should be done and why it should be done to build and sustain integrated supply chains for the post-COVID era of high uncertainty and multiple risks.

Next, the third category showcases the state-of-the-art molding-based six innovative solutions for industry-specific (e.g., healthcare, transport, livestock and poultry, pharmaceutical, and automotive) supply chains including *Sustainable Outcomes Through The Structured Forward Supply Chain: A System Dynamic Approach, Using System Dynamics modeling to frame environmental voluntary commitment programs in the transport industry in the post-COVID situation: the French experience, Waste Processing Scenarios to Support Sustainable Environmental Development Using*

*System Dynamics, Multi-criteria analysis of disruption risks for supply chains due to pandemics, Botswana state-owned enterprises (SOE): A review of lessons learned from supply chain management best practices during COVID-19, and Collaboration Model between Buyer and Supplier: An Empirical Assessment of Indonesian Pharmaceutical Industr*y. These applications address the critical question of "how to do" by demonstrating the design, development, and application of modeling-based innovative solutions aimed at improving the performance of supply chains in the post-COVID era. Finally, the last category lists the thematic bibliography to advance future research about post-COVID supply chains.

The unique feature of this book is that each of the eight chapters explicitly presents a section about the ***implications for Post-COVID supply chains practice and theory***. These implications are specifically are presented as viable insights for managing and coping with the risks and uncertainties due to unprecedented and unpredictable effects of external shocks like COVID-19.

It is worth noting that although the model-based contributions in this volume have been applied to only five specific industry-related issues, the insights presented and the model structures developed and presented in this book are generic enough to be applied to the supply chains of other domains e.g., energy, aviation, and retail industry. Likewise, several chapters in this book have presented "future research opportunities" for researchers in the domain of supply chains to avail.

References

Barbieri, P., Bofelli, A., Elia, S., Fratocchi, L., Kalchschmidt, M., & Samson, D. (2020). What can we learn about reshoring after Covid-19? *Operations Management Research, 13*(3), 131–136.

Kiers, J., Seinhorst, J., Zwanenburg, M., & Stek, K. (2022). Which Strategies and corresponding competencies are needed to improve supply chain resilience: A COVID-19 based review. *Logistics, 6*(12), 1–17.

Paul, S. K., Moktadir, M. A., & Ahsan, K. (2021). "Key supply chain strategies for the post-COVID-19 era: Implications for resilience and sustainability", *The International Journal of Logistics Management*, Vol. ahead-of-print No. ahead-of-print. https://doi.org/10.1108/IJLM-04-2021-0238.

Sajjad, A. (2021). The COVID-19 pandemic, social sustainability, and global supply chain resilience: A review. *Corporate Governance, 21*(2), 1–17.

Part II
Understanding Supply Chain Dynamics -Theoretical Perspectives

Chapter 2
Post-Covid Supply Chain Dimensions Proposed Shift

Fabián Szulanski and Hassan Qudrat-Ullah

Abstract Covid 19 pandemics has demonstrated being a Black Swan, which is a non-expected event with a disruptive impact, forcing the world to change most of the former ways of value creation, collaboration, strategizing and decision making. Supply chain management and the logistics organizational function have also been impacted by this fierce disruption. The author shares some observations of what happened with supply chain decision making during the lockdown, and after that, he proposes several shifts in how strategy is designed, in how decisions are made, and in how should collaboration happen in the supply chain ecosystem. In all instances he describes the expected impacts on post-Covid supply chain decision making, expecting that the proposed shifts could be used by researchers, practitioners, and decision makers to propel further research and strategic experimentation.

Keywords Supply chain · Post Covid · Disruption · Adaptation · Shifts · Decision making · Cultural transformation · Internal processes attunement · Customer-centric · Value flows · Value network · Customer satisfaction · Business ecosystem · Experiential service logic · Logistics · Value field · Transformation · Emergent phenomena · Living organizations · Ephemeral teams · Self-organization · Decentralization

2.1 Introduction

In this book chapter, the authors will share some observed supply chain transformations in the Post Covid context. Additionally, he will propose some supply chain dimension shifts to offer possibilities to strengthen leaders' competencies and

F. Szulanski (✉)
Universidad del CEMA, Buenos Aires, Argentina
e-mail: fhszulansk@ucema.edu.ar

H. Qudrat-Ullah
School of Administrative Studies, York University, Toronto, ON, Canada
e-mail: hassanq@yorku.ca

H. Qudrat-Ullah (ed.), *Understanding the Dynamics of New Normal for Supply Chains*, Understanding Complex Systems,
https://doi.org/10.1007/978-3-031-07333-5_2

preparedness to tackle those transformations, within, in transition from, and letting go of the Industrial mindset.

Some of the observed Post Covid transformations in the supply chain were in the end customer demand, in back-office communication and collaboration, in organizational design, and in cultural aspects that had their impact on logistics strategic, operational, and tactical decision making aspects.

The proposed Post Covid shifts that will be covered in this chapter are around how the organization (and thus the supply chain function) may be considered as a living organism rather than as a mechanism, how to enable leverage in the logistics function's reputation, how to transcend valuing logistics just as a technical process and consider socio-technical and perceptual aspects, how to focus on present moment needs and letting go of planning and foresight. The authors will also analyze how each of those Post Covid proposed shifts may impact the supply chain decision making process.

2.2 Observed Post Covid Transformation in Supply Chain

2.2.1 Last Mile Service

Customers want their merchandise to arrive faster. Srinivas and Marathe (2021) introduce the idea of a mobile warehouse, allocating a truck for a certain neighborhood of zip code zone as an efficient way of tackling the last mile delivery issue in post-Covid times, while also doing a review of other solutions. Today's digital businesses heavily rely on reliable, efficient, and prompt logistics operations.

Mental health issues such as anxiety, stress, loneliness abounded during Covid lockout phases (Hiremath et al., 2020; Pieh et al., 2021; Rossi et al., 2020). The retail business ecosystem was no exception. Roggeveen and Sethuraman (2020) expand on the different behaviors and expectations post-Covid and highlight the challenges suppliers need to address to satisfy the new normal needs. Examples are how to supply at an appropriate time of the day and with minimal errors in delivery consolidation as consumers are working mostly from home and buying more in retail e-commerce online outlets. Other challenges involve generating impulse buying in technology-mediated environments. The most pervasive impact related to commerce is that buyers' patience level has been decreased over time, having expectations of sustainingly decreasing waiting times for receiving goods and services they purchased. Therefore, competitors in different industry verticals, especially in the business-to-consumer business model, struggle to cut corners in their logistic processes, to shorten the duration of their service's last mile.

What systemic impact could this supply chain process reengineering have on different aspects of the business? The following causal loop diagram may help make sense of this (Fig. 2.1):

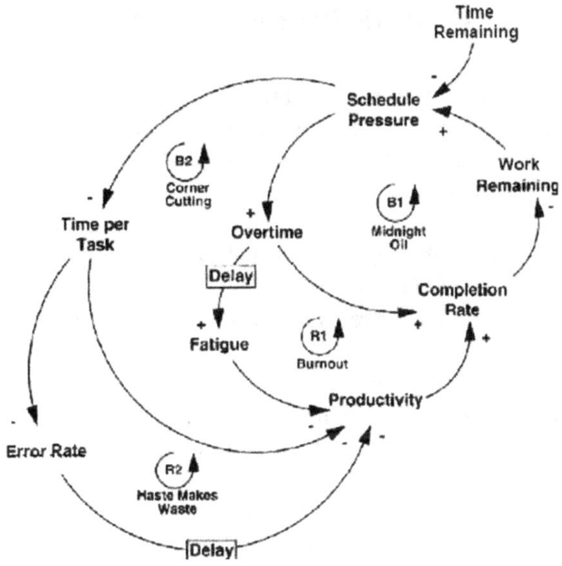

Fig. 2.1 Cutting corners [Salem (2018), based on Sterman (2000)] (*Source* Salem [2018] reformatted from Sterman [2000])

Reduction of time per task increases productivity but increases overtime work, and just after a short delay, it will trigger fatigue, which in turn, will negatively impact on employee's motivation, which will increase error rate and therefore decrease productivity. It is a Fixes that Fail systems archetype (Kim, 1995), in which the most sustainable intervention is to discover the action (not a fix) that triggers an unintended consequence with a risk that is below the tolerance threshold.

Impact on decision making: Rather than stressing the supply chain with overtime and cutting corners, with the aforementioned negative impact, companies may choose other incentives for buyers so that they don't fall into this systemic trap and lose repeated purchases from recurrent customers. They could incentivize patience with some additional service perks in all touchpoints of customers' experience. Szulanski and Viñegla (2003) propose that customer satisfaction should stretch towards the delight of the value network of the supplier, and in all touchpoints. Sharma et al. (2020) argues that collabaorative decsion making with the parterns of a supply chain network can improve trust among the network parterns. As a result, collaborative approach can lead to data sharing and joint problem solving among the partners resulting in the reduction of supply chain disruptions and risks. Then, the supplier may oscillate its resource allocation between the polarities of reducing delivery times up to a level that could be sustained, and offering service perks, as Johnson (1993) proposed in his polarity management model. Thresholds for switching to the other polarity are determined by an early alert system.

2.2.2 Dramatic Increase in Remote Work

Throughout Covid times, remote work has been a required modality. Brynjolfsson et al. (2020) expand on this phenomenon by sharing their findings: that n the face of external shocks like COVID-19, (i) more people were working remotely, and (ii) states with a higher share of employment in IT-related workforce were more likely embrace remore work. Its adoption time has shortened, even for late adopters, because of the rapid contextual change catalyzed by Covid and the related long lockout periods.

People who were averse to technology, who preferred live, in-office meetings, were then almost forced to engage in nonstop technology-mediated conversations, also impacting their cognitive, physical, and mental health, draining even more energy and redefining the present of how work was supposed to be executed. Organizations may create and hold a space for nurturing and treating those who are averse to technology utilization. Wheatley and Frieze (2011) propose a two-loop model that includes such a space, which they name as the Hospice. There were controversies regarding the use of cameras and microphones during those meetings. Karl et al. (2021) expand on some of the excessive video conference platform usage related challenges, such as mental fatigue, excessive direct eye contact time, stress accumulation; also sharing insights from a Gartner's study saying that just 25% of meetings post-Covid will be in person by 2024 (Standaert et al., 2021).

Impact on decision making: Overstressed employees in all corporate functions, had to make decisions with their cognitive sharpness being decreased. Therefore, all phases of decision making were affected negatively. Reactive rather than adaptive (Ali et al., 2021), tactical rather than strategic (Kumar & Sharma, 2021), those attributes reflected the type of policy design (if any) and decision making process during times of remote work. Team building and trust-building in supply chain management were also affected during Covid (Khudhair et al., 2020) because of the dominance of remote work, additionally affecting collective agreements and decision making processes from the different teams composing the supply chain function.

2.3 Organizational Design Dimension

During Covid times, organizations have delegated some agency to teams, adopting flat organizational design (Fu et al., 2020). Ness also considers long-term effects regarding organizational design post-Covid and posits that (i) organizational tasks that are higly interdependent will tend to become more modular, and (ii) the interface with work activies will need rethinking. The supply chain function has been impacted because the installation of new interaction protocols was enacted without preparing beforehand the human beings comprising the supply chain teams for that change.

Post-Covid impact on decision making was that people didn't adopt the new ways of interacting and co-creating value, expecting top-down directives as it happened during pre-Covid times. They were not prepared to enjoy the good and the bad of being free agents within self-determined, self-organized teams. The authors propose a shift to consider organizations as living organisms. What does that imply, and what impact does it have in supply chain decision making? A living organism is autonomous (Franklin & Graesser, 1996), decentralized (Christie et al., 2003), distributed (Jia & Zhou, 2004) and self-organized. It doesn't necessarily have a permanent governance body, for example in the case of the single-cell blob described in (Tennenhouse, 2017). How would an organization function following this paradigm? It would be composed of ephemeral, emergent present moment needs centered teams, which would form and dismantle as soon as those needs are satisfied, as described in Gray and Vander Wal (2014). The purpose would be defined by a temporary assembly such as a Board of Directors, but right after the purpose is defined, that assembly would dismantle. The organization, then, will define deep values and ways of working, collectively, and function through self-organization thereafter.

Impact on supply chain management decision making: Decision making will be closer to the emergent need, therefore more accurate. The decision making process needs to become adaptive and agile to be available even if teams continuously form and dismantle.

Impacts of that ephemerality could be alleviated by helping make decisions with technology, through the establishment of a DAO (Decentralized Autonomous Organization) based on a Blockchain platform.

2.3.1 *Cultural Aspects: Lack of 'Socio' in Socio-Technical*

A cultural focus on efficiency and productivity in supply chain management, which translated into triggering acceleration in the speed of work during Covid (Hitt et al., 2021) has impacted many dimensions in decision making, reinforcing those created by organizational design in the above section.

Alleviating those impacts requires many cultural shifts: from a focus on results to a focus on human beings within the business ecosystem, including internal collaborators, as developed in (Szulanski & Viñegla, 2003). Plus nurturing a mindset shift regarding decision making in supply chain management towards sustainability (Rajeev et al., 2017), a decrease in energy consumption (Wee et al., 2012), and decarbonization (Brinken et al., 2022).

Impact in supply chain management decision making is that every action or communication should pass all those filters before they are enabled to be enlivened and enacted, guaranteeing that all new actions or expressions will be values aligned and functioning in the new paradigm being proposed (Hales & Pronovost, 2006). However, some paradoxes may generate tensions, which should be managed (Brix-Asala et al., 2018).

2.3.2 *Reputation and Trust: From Customer Satisfaction to Business Ecosystem Experience*

2.3.2.1 NFT Tokenization

Traditional organizations are most focused on pushing their products into their target markets. At best, some forward-thinking organizations are customer-centric. Is it enough? (Komatsu Cipriani et al., 2020). The author's opinion is that it's not. As Szulanski and Viñegla (2003) propose, all nodes of a business ecosystem should feel delighted by what they experienced when being impacted by the value offer of the supplying company.

On the other hand, the author believes that trust in other nodes of the business ecosystem should be objective and transparent. New technologies, such as blockchain and related applications such as NFTs, are allowing trust objectivization and transparency (Azzi et al., 2019; Francisco & Swanson, 2018; Nadini et al., 2021; Queiroz et al., 2019). However, as Mazzei et al. (2020) share their findings that (i) there is a challenge in linking industrial assets and Blockchain because it can lead to cost and technological issues for the firms, and (ii) the interoperability, portability, scalability, and security issues related to the adoption of Blockchain technogies.

The supply chain function offers many experiential touchpoints to the corporate and consumer ecosystem (Rao et al., 2021), thus having the opportunity of redefining how reputation and trust are taken into account (Benton et al., 2018; Shahzad et al., 2021). This is most important after Covid, as business ecosystem reputation and trust are volatile and very sensitive to efficiency (Obrenovic et al., 2020). Calcaterra believes that DAOs (Distributed Autonomous Organizations) is the best way to overcome this challenge (Calcaterra & Kaal, 2020). Supply chains may adopt a governance system similar to that of a DAO (Lohmer et al., 2021).

The impact of adopting a more adaptive and decentralized stance for supply chain decision making, is the decrease in office politics pressures, misalignments, while gently nudging contributors to behave in a trustful way, improving the reputation of the supply chain function, and of the corporation in general, adding to an increase in its brand value, as stated in (Smith et al., 2010). This catalyzed context will offer a more tranquil environment for decision making in general and in supply chain management in particular, as is described in (Swan, 2015), *Focusing on the present moment: emergent conversations in supply chain management.*

This proposed shift is related to letting go of the planning and adopting the sense and respond adaptive organizational behavior (Haeckel, 1995), which is better suited for dealing with complex issues and for performing in complex contexts, from which the logistics function and supply chain management cannot escape. Kurtz and Snowden (2003) expand on the usefulness of the adaptive stance for performing in complex situations. David Whyte (2002) says "conversation is the work". The author deeply agrees with this, believing that an organization can be seen as three intertwined networks: conversations, meaning, and value.

On the other hand, Greek philosopher Hieraclitus said "No man ever steps in the same river twice, for it's not the same river and he's not the same man", pointing that only the present moment is relevant for making sense of what is happening. Extrapolation of past events (Lakonishok et al., 1994), and foresight of uncertain happenings (Nyce & Cpcu, 2007), may imply significant errors in decision making. Integrating the aforementioned insights, the author proposes a shift: from planning, towards holding spaces for emergent conversations.

Some may object to the holding space meme. However, is it possible to manage emergent conversations? Desai et al. (2021) expand on why is it needed to hold space for emergent insights so that narratives could be catalyzed. Eden et al. (2021) expand on emergent strategizing and improvisation, which fit this bill.

What's the benefit of supply chain decision making? Dealing with the present moment, sense and adapt to current circumstances, being flexible enough to rearrange value networks and resource allocation, allows for a more attuned and synchronized way for performing in supply chain's complex context, in which human being perceptions may bias decisions and create emergent impacts that ask for a rapid adaptation and change of course.

2.4 Key Implications for Post Covid Supply Chain Practitioners

Before concluding, the key implications for Post Covid Supply Chain Practitioners, after all, what has been observed and suggested, can be categorized in the following categories, regarding how their proactive is going to shift.

2.4.1 Technological

Practitioners will have to know what emergent and future technologies they can curate and choose for integrating and making the supply chain ecosystem's experience thrive. They will have to be good translators from logistics demand to technological suppliers, for explaining functional requirements from the former to the latter.

2.4.2 Competences

Practitioners will have to master community building and space holding capacities, as conversational communities of practice are going to predominate from the near future. This involves them learning through action how to facilitate, hold space, moderate, coach, and catalyze those conversational spaces. As per their functional leadership, they will have to learn efficient and inspiring ways of developing attractive narratives, learning transformative storytelling techniques. Efficiency: For dealing with instant gratification expectations, they will have to be experts in simplifying processes, eliminating unnecessary bureaucracies, and working in small, collaborative, ephemeral teams. They will have to master how to co-create with practitioners from other business functions, transitioning into being generalists with deep expertise in supply chain managerial decision making.

2.4.3 Attitudinal

Flexibility, Adaptability, Tolerance to Error, Proactivity, Psychological endurance, and Resilience for working on many projects at the same time, will be the main traits the supply chain practitioner should possess or develop. Other needed traits are transparency, vulnerability, willingness to learn, emotional and relational intelligence, courage, and openness to uncertainty.

2.5 Conclusion

New times call for a shift in traditional supply chain management decision making approaches. More specifically, supply chain management hasn't been immune to current happenings related to Covid 19 impacts. Several aspects of those impacts have been observed. The author has analysed and criticised traditional planning and predictive forecasting practices in supply chain managerial decision making. The author has also proposed some shifts in different dimensions of post-Covid supply chain management and decision making, many of them realizing the value of considering human beings as relevant stakeholders in that apparent sterile process. The author expects that this brief work could inspire researchers and practitioners so that they may experiment adopting some or all the proposed shifts, and generate further research and experimentation.

References

Ali, M. H., Suleiman, N., Khalid, N., Tan, K. H., Tseng, M. L., & Kumar, M. (2021). Supply chain resilience reactive strategies for food SMEs in coping to COVID-19 crisis. *Trends in Food Science & Technology, 109*, 94–102.

Azzi, R., Chamoun, R. K., & Sokhn, M. (2019). The power of a blockchain-based supply chain. *Computers & Industrial Engineering, 135*, 582–592.

Benton, M. C., Radziwill, N. M., Purritano, A. W., & Gerhart, C. J. (2018). Blockchain for supply chain: Improving transparency and efficiency simultaneously. *Software Quality Professional, 20*(3).

Brinken, J., Pabsch, C., & Behrendt, F. (2022). Decarbonization potential of logistic 4.0 technologies in apple supply chains. *Procedia Computer Science, 200*, 461–470.

Brix-Asala, C., Geisbüsch, A. K., Sauer, P. C., Schöpflin, P., & Zehendner, A. (2018). Sustainability tensions in supply chains: A case study of paradoxes and their management. *Sustainability, 10*(2), 424.

Brynjolfsson, E., Horton, J. J., Ozimek, A., Rock, D., Sharma, G., & TuYe, H. Y. (2020). *COVID-19 and remote work: An early look at US data* (No. w27344). National Bureau of Economic Research.

Calcaterra, C., & Kaal, W. A. (2020). Reputation protocol for the internet of trust. In *Legal Tech and the New Sharing Economy* (pp. 155–179). Springer.

Christie, A. A., Joye, M. P., & Watts, R. L. (2003). Decentralization of the firm: Theory and evidence. *Journal of Corporate Finance, 9*(1), 3–36.

Desai, A., Nguyen, H. N., & Nagda, B. R. A. (2021). Honoring culture, holding complexity: Synthesis and emerging possibilities in dialogue. In *Global Perspectives on Dialogue in the Classroom* (pp. 175–199). Palgrave Macmillan.

Eden, C., Ackermann, F., & Vito, V. (2021, June). Improvisation and emergent strategizing: The role of group support systems. In *International Conference on Group Decision and Negotiation* (pp. 16–24). Springer.

Foss, Nicolai J. (2020). "The impact of the Covid-19 pandemic on firms' organizational designs". *Journal of Management Studies.*

Francisco, K., & Swanson, D. (2018). The supply chain has no clothes: Technology adoption of blockchain for supply chain transparency. *Logistics, 2*(1), 2.

Franklin, S., & Graesser, A. (1996, August). Is it an agent, or just a program? A taxonomy for autonomous agents. In *International workshop on agent theories, architectures, and languages* (pp. 21–35). Springer.

Fu, H., Zhao, C., Cheng, C., & Ma, H. (2020). Blockchain-based agri-food supply chain management: Case study in China. *International Food and Agribusiness Management Review, 23*(5), 667–679.

Gray, D., & Vander Wal, T. (2014). *The connected company.* "O'Reilly Media, Inc."

Haeckel, S. H. (1995). Adaptive enterprise design: The sense-and-respond model. *Planning Review.*

Hales, B. M., & Pronovost, P. J. (2006). The checklist—a tool for error management and performance improvement. *Journal of Critical Care, 21*(3), 231–235.

Hiremath, P., Kowshik, C. S., Manjunath, M., & Shettar, M. (2020). COVID 19: Impact of lock-down on mental health and tips to overcome. *Asian Journal of Psychiatry, 51*, 102088.

Hitt, M. A., Holmes, R. M., Jr., & Arregle, J. L. (2021). The (COVID-19) pandemic and the new world (dis) order. *Journal of World Business, 56*(4), 101210.

Jia, W., & Zhou, W. (2004). *Distributed network systems: From concepts to implementations* (Vol. 15). Springer Science & Business Media.

Johnson, B. (1993). Polarity management. *Executive Development, 6*, 28–28.

Karl, K. A., Peluchette, J. V., & Aghakhani, N. (2021). Virtual work meetings during the COVID-19 pandemic: The good, bad, and ugly. *Small Group Research*, https://doi.org/10.1177/104649642 11015286

Kim, D. H. (1995). Systems archetypes as dynamic theories. *The Systems Thinker, 6*(5), 6–9.

Khudhair, H. Y., Alsaud, A. B., Alsharm, A., Alkaabi, A., & AlAdeedi, A. (2020). The impact of COVID-19 on supply chain and human resource management practices and future marketing. *International Journal Supply Chain Management, 9*(5), 1681.

Komatsu Cipriani, T., Deserti, A., Kleverbeck, M., Rizzo, F., & Terstriep, J. (2020). Business models & social innovation: Mission-driven versus profit-driven organisations. *International Review of Applied Economics, 34*(5), 541–566.

Kumar, B., & Sharma, A. (2021). Managing the supply chain during disruptions: Developing a framework for decision-making. *Industrial Marketing Management, 97*, 159–172.

Kurtz, C. F., & Snowden, D. J. (2003). The new dynamics of strategy: Sense-making in a complex and complicated world. *IBM Systems Journal, 42*(3), 462–483.

Lakonishok, J., Shleifer, A., & Vishny, R. W. (1994). Contrarian investment, extrapolation, and risk. *The Journal of Finance, 49*(5), 1541–1578.

Lohmer, J., Petzok, L., & Lasch, R. (2021). Governance design of blockchain consortia for efficient and transparent procurement and supply chain management. In *Supply Management Research* (pp. 117–134). Springer Gabler.

Mazzei, D., Baldi, G., Fantoni, G., Montelisciani, G., Pitasi, A., Ricci, L., & Rizzello, L. (2020). A Blockchain Tokenizer for Industrial IOT trustless applications. *Future Generation Computer Systems, 105*, 432–445.

Nadini, M., Alessandretti, L., Di Giacinto, F., Martino, M., Aiello, L. M., & Baronchelli, A. (2021). Mapping the NFT revolution: Market trends, trade networks, and visual features. *Scientific Reports, 11*(1), 1–11.

Nyce, C., & Cpcu, A. (2007). Predictive analytics white paper. *American Institute for CPCU. Insurance Institute of America*, 9–10.

Obrenovic, B., Du, J., Godinic, D., Tsoy, D., Khan, M. A. S., & Jakhongirov, I. (2020). Sustaining enterprise operations and productivity during the COVID-19 pandemic: "Enterprise Effectiveness and Sustainability Model." *Sustainability, 12*(15), 5981.

Pieh, C., Budimir, S., Delgadillo, J., Barkham, M., Fontaine, J. R., & Probst, T. (2021). Mental health during COVID-19 lockdown in the United Kingdom. *Psychosomatic Medicine, 83*(4), 328–337.

Queiroz, M. M., Telles, R., & Bonilla, S. H. (2019). Blockchain and supply chain management integration: A systematic review of the literature. *Supply Chain Management: An International Journal.*

Rajeev, A., Pati, R. K., Padhi, S. S., & Govindan, K. (2017). Evolution of sustainability in supply chain management: A literature review. *Journal of Cleaner Production, 162*, 299–314.

Rao, S., Gulley, A., Russell, M., & Patton, J. (2021). On the quest for supply chain transparency through blockchain: Lessons learned from two serialized data projects. *Journal of Business Logistics, 42*(1), 88–100.

Roggeveen, A. L., & Sethuraman, R. (2020). How the COVID-19 pandemic may change the world of retailing. *Journal of Retailing, 96*(2), 169.

Rossi, R., Socci, V., Talevi, D., Mensi, S., Niolu, C., Pacitti, F., … & Di Lorenzo, G. (2020). COVID-19 pandemic and lockdown measures impact on mental health among the general population in Italy. *Frontiers in Psychiatry*, 790.

Salem, T. J. (2018). *An Investigation of key risks and risk management strategies in construction projects-Gaza Strip.* MSc. in Civil Engineering Thesis, The Islamic University-Gaza.

Shahzad, A., Wenyu, C., & Kumar, R. (2021, April). Blockchain based monitoring on trustless supply chain processes. In *2021 IEEE 6th International Conference on Cloud Computing and Big Data Analytics (ICCCBDA)* (pp. 216–221). IEEE.

Sharma, A., Adhikary, A., & Borah, S. B. (2020). Covid-19's impact on supply chain decisions: Strategic insights from NASDAQ 100 firms using Twitter data. *Journal of Business Research, 117*, 443–449.

Smith, K. T., Smith, M., & Wang, K. (2010). Does brand management of corporate reputation translate into higher market value? *Journal of Strategic Marketing, 18*(3), 201–221.

Srinivas, S. S., & Marathe, R. R. (2021). Moving towards "mobile warehouse": Last-mile logistics during COVID-19 and beyond. *Transportation Research Interdisciplinary Perspectives, 10*, 100339.

Standaert, W., Muylle, S., & Basu, A. (2021). How shall we meet? Understanding the importance of meeting mode capabilities for different meeting objectives. *Information & Management, 58*(1), 103393.

Sterman, J. (2000). *Business dynamics: Systems thinking and modeling for a complex world* (p. 564). Irwin/McGraw-Hill.

Swan, M. (2015). *Blockchain: Blueprint for a new economy.* O'Reilly Media, Inc.

Szulanski, F., & Viñegla, A. L. (2003). XSC o balanced scorecard basado en laexperiencia y las vivencias de la red de valor. *Estrategia Financiera, 194*, 38–44.

Tennenhouse, E. (2017). It's a no-brainer. *New Scientist, 235*(3134), 32–35.
Wee, H. M., Yang, W. H., Chou, C. W., & Padilan, M. V. (2012). Renewable energy supply chains, performance, application barriers, and strategies for further development. *Renewable and Sustainable Energy Reviews, 16*(8), 5451–5465.
Wheatley, M., & Frieze, D. (2011). *Walkout, walk on.* Berrett-Koehler Publishers.
Whyte, D. (2002). *Crossing the unknown sea.* Riverhead Books.

Chapter 3
Characteristics and Capabilities of a Successful Supplier: A Conceptual Model

Hassan Qudrat-Ullah and Fabián Szulanski

3.1 Introduction

Increasing local competition, globalization, and technological advancements, uncertain and dynamic business environments, and increasing focus on green and sustainable production and consumption of goods and services require businesses to build and sustain effective and efficient supply chains. The successful performance of any supply chain heavily depends on how good its suppliers are. For instance, selecting suitable suppliers significantly reduces material purchasing costs, improves the competitiveness of businesses, increases flexibility and product quality, and helps with speeding up the process of material purchasing for both the manufacturing and service industry (Sajjad, 2021; Xia & Wu, 2007). The current onslaught of pandemic (i.e., COVID-19) on businesses and their supply chains have made the role of suppliers even more critical and pronounced. Building strong and coordinated relationships with suppliers has become a strategic decision for the firms. With strong partnerships with suppliers, firms can share resources and manage overall supply chain risks (Kiers et al., 2022). For instance, as empirical evidence to the importance of the selection of the suppliers, Dell company survived the economic slowdown in 2001 and increased their sales by 15% while other competitors dropped 5% thanks to their strategy and supplier's relationship (Kiers et al., 2022). It is, therefore, critical to for the management of firms to focus on the modalities and criteria for the selection of suppliers.

The traditional focus on "cost competitiveness" alone is hardly enough in the selection of suppliers. When "suppliers" are considered partners, what should be the

H. Qudrat-Ullah (✉)
School of Administrative Studies, York University, Toronto, ON, Canada
e-mail: hassanq@yorku.ca

F. Szulanski
UCEMA University, Buenos Aires, Argentina
e-mail: fhszulansk@ucema.edu.ar

H. Qudrat-Ullah (ed.), *Understanding the Dynamics of New Normal for Supply Chains*, Understanding Complex Systems,
https://doi.org/10.1007/978-3-031-07333-5_3

key characteristics and capabilities of potential suppliers. The purpose of this Chapter, therefore, is to review research studies about supplier selection criteria in an attempt to determine the key characteristics and capabilities of suppliers for the development of long-term relationships between manufacturers (and service providers) and suppliers. Specifically, we will attempt to identify key characteristics of a first-tier supplier in the automotive industry. Our results show that price, quality, and delivery remain important factors but communication skills, information technology, human capital, and strategic commitment are increasingly becoming vital factors for the supplier selecting process. Our findings and insights are useful both for suppliers and manufacturers (and service producers) who are interested in developing and using these relationships to overcome the challenges of an ongoing uncertain and dynamic business environment.

The next section presents a critical review of literature about supplier selection.

3.2 Supply Chain Management in the Twenty-First Century

3.2.1 History and Development

The concept of the supply chain in management could be traced back to the early twentieth century. During this time, with the introduction of the assembly line, the concept of the supply chain became great important. However, not until 1982, the term "supply chain management" became popular when it was first used by Keith Oliver.

There are several definitions of supply chain management provided by scholars. Supply chain management is the systematic, strategic coordination of the traditional business functions and the tactics across these business functions within a particular company and across businesses within the supply chain, to improve the long-term performance of the individual companies and the supply chain as a whole (Mentzer et al., 2001).

A customer-focused definition is given by Hines: "Supply chain strategies require a total systems view of the linkages in the chain that work together efficiently to create customer satisfaction at the endpoint of delivery to the consumer. As a consequence costs must be lowered throughout the chain by driving out unnecessary costs and focusing attention on adding value. Throughput efficiency must be increased, bottlenecks removed and performance measurement must focus on total systems efficiency and equitable reward distribution to those in the supply chain adding value. The supply chain system must be responsive to customer requirements".

In summary, supply chain management is the umbrella that covers all aspects of the sourcing and procurement of goods. Supply chain management forms and manages the business-to-business links that allow for the ultimate sale of goods to consumers.

Throughout the history of the development of supply chain management studies, one could observe three major movements which included: Creation, Integration, and Globalization (Movahedi et al., 2009). Now with COVID-19's unprecedented and unpredictable impact on supply chains, it is not unreasonable to say that we are now at the beginning of a Post-COVID ear for suppl chains.

3.2.2 Creation Era

As mentioned earlier, the term "Supply Chain Management" was first used in the 1980s and as such is a relatively new discipline within a management theory about tools and concepts still being developed (Lummus & Vokurka, 1999).

The characteristics of the early era of supply chain management included the need for radical changes in operations such as re-engineering and downsizing driven by the cost reduction programs and the Japanese practice of management.

3.2.3 Integration Era

The use of integrated systems in supply chain management thrived during the 1960s to the 1990s when firms started using systems such as Electronic Data Interchange (EDI) and Enterprise Resource Planning (ERP) to increase value-adding and cost reductions through integration.

One could classify the supply chain into 3 stages. In stage 1, data was stored in various systems such as Storage, Material, and Distribution. These systems were not linked and independent of each other. In stage 2, these systems were linked together with the introduction of Enterprise Resource Planning (ERP) systems. In stage 3, vertical integrations in both upstream with suppliers and downstream with customers became the trend. Well-integrated firms and supply chains partners including suppliers appear to succeed in managing supply chain risks and uncertainties (Sajjad, 2021).

3.2.4 Globalization Era

The last movement of supply chain management development, the globalization era, can be characterized by the attention given to global systems of supplier relationships and the expansion of supply chains over national boundaries and into other continents. Although the use of global sources in the supply chain of organizations can be traced back several decades, it was not until the late 1980s that a considerable number of organizations started to integrate global sources into their core business. This era is characterized by the globalization of supply chain management in organizations to increase their competitive advantage, value-adding, and reduce costs through global sourcing.

3.2.5 Role of a Supplier in the Supply Chain

In the current uncertain and dynamic environments, the supply chain has become the central organizing unit in today's global industries. The competition is among supply chain networks rather than individual firms. Firms are challenged by the need to effectively manage increasingly extending supply chain activities beyond the boundary of the firms (Lee et al., 2001). Firms invest in the supply chain system to share and collaborate between suppliers, dealers, customers, and partners. In a collaborative supply chain environment, the supply chain members work together, share important information, and collaborate on activities efficiently and effectively.

Given the importance of each member in the supply chain, firms nowadays pay close attention to the process of selecting their suppliers. Buyers evaluate their future partners through a set of "supplier selection criteria" before conducting business. Generally, buyers have particular qualifiers and order winners, which form the grounds on which buyers select and co-operate with suppliers and judge their performance. According to Terry (2002), an order qualifier is a characteristic of a product or service that is required for the product/service to even be considered by a customer. An order winner is a characteristic that will win the bid or customer's purchase. Hill and Hill (2009) suggested that suppliers should verify the importance of buyers' order-winners by analyzing data from actual customer orders. Thus, there is a need for suppliers must identify these qualifiers and order winners, and know the criteria on which they will be judged.

As a result, the purpose of this paper is to review different studies about supplier selection criteria to find out the key characteristics and capabilities needed to be a successful supplier in the twenty-first century.

3.3 Supplier Selection Criteria

The supplier selection process has undergone significant changes during the past three decades. These include increased quality guidelines, improved computer communications, and increased technical capabilities (Weber et al., 1991). Buyer-supplier relationships based solely on price are no longer acceptable for suppliers of critical materials or for organizations that wish to practice the latest innovations in supply chain management. Recent emphasis has also been on other important strategic and operational factors such as quality, delivery, and flexibility. Strategic relationships also play a vital role in the long-term well-being of a supply chain (Sarkis & Talluri, 2002).

Shuyong and Rongqiu (1998) maintained that supplier evaluation should rely on the following attributes: quality, delivery period, batch flexibility, the balance between the delivery period and price, the balance between the price and batch, variety, etc. Shihua and Xubin (2002) developed an integrated evaluating attribute system for selecting co-partners under the circumstances of supply chain management and generalized four main factors that could affect co-partner selection: outstanding achievement of an enterprise, operation structure and throughput, quality system, and enterprise environment. Lijuan (2002) proposed that criteria for supplier selection were composed of nine evaluating attributes: product quality, product price, after service, distance, technological level, supply capability, economic revenue, delivery, and market influence. In the construction industry, suppliers offer heavy equipment and machinery, labor, building materials, service expertise, etc. (Florez-Lopez, 2007; Lam et al., 2010; Ustun & Demirtas, 2008).

Based upon these literature reviews, the supplier selection criteria could be categorized into two separate categories: Tangible and Intangible. In some journal articles, the set of Tangible criteria is referred to as Hard factors and the set of Intangible criteria is referred to as Soft factors. Quantifiable or "hard" criteria (Ellram, 1990) such as price, delivery, quality, and service are routinely used for supplier selection and assessment. "Soft," difficult-to-quantify factors such as management compatibility and strategic direction of the supplier have also been shown to be important, particularly in the context of strategic buyer-supplier partnerships (Ellram, 1990).

Each category contains different sets of characteristics that a buyer would look for in a supplier (Appendix 1).

The tangible criteria include:

- Price /cost
- Product quality
    1. The rejection rate of the product
    2. Increase lead time
    3. Quality assessment
    4. Remedy for quality problems
- Service performance
    1. Delivery reliability and responsiveness
    2. Technological and R&D support
    3. Response to changes
- Human capital
- Production facility and capacity

The intangible criteria include:

- Reputation
- Trust /Honesty
- Commitment
- Performance history

3.4 Importance of Supplier Selection Criteria

In this following section, the paper will discuss the importance of the criteria and explain why purchasing firms are looking for those characteristics in a supplier.

3.4.1 Overall Price/Cost Performance

This criterion is one of the important criteria in assessing the supplier because it can directly influence the acquisition of goods or services cost. Profit maximization can be achieved with cost minimization. Thus, the buying firm always requires the minimum price of the product to increase its profitability. As a result, they are always looking for a low-cost supply base where they can minimize manufacturing costs related to the production of the product. The processing cost, maintenance cost, warranty cost, and other costs related to the manufacturing of the product determine the total price of the product (Kaplan, 1998).

3.4.2 Quality of the Product

Another important factor that the purchasing firm requires is the high quality of the products. The quality of the product can be measured in terms of the following attributes:

The rejection rate of the product: The rejection rate of the product is defined in the terms of the number of parts rejected by the customers in a fixed period because of some quality problems (Lee, 2003, p. 229).

Increased lead time: The defective parts, which are not detected in the quality control process but noticed during production, can increase the lead time of production (Lee, 2003, p. 228).

Quality assessment: The buyer would investigate whether or not potential suppliers are certified for strict quality assurance and have a strong commitment to preventing quality failures (Chao, 2009).

Remedy for quality problems: The purchasing firm is also interested in the ability of the supplier to efficiently tackle the quality problems (Balachandran, 2005).

3.4.3 Service Performance

The performance of the supplier in providing service to the manufacturer is the prime criteria to decide its suitability for a particular product. The good service given by the supplier may help in increasing the customer base and therefore, this criterion is important in supplier selection. It is analyzed based on the following attributes:

Delivery reliability and responsiveness: The ability of the supplier to supply the customer demand on time has always been the critical criterion for selection in this dynamic environment (Gorchels, 2004, p. 71).

Technological and R&D support: In this fast-moving world, technology is advancing at a very fast pace. In addition, suppliers are more likely to assume greater responsibility for outsourced design, engineering service, prototype development, and research to increase the performance of the products. The suppliers' ability to provide advanced technological and R&D support to produce a good product is of prime concern in supplier selection.

Response to changes: The ability of the supplier to change according to the buyer's demand, price structure, order frequency, and current business scenario can affect the performance of the firm in case of urgent and uncertain demand. As a result, suppliers with flexibility can be chosen for better performance towards the customer (Gorchels, 2004, p. 71).

Human Resource (Staff): This criterion is tied directly with communication skills. Suppliers' salesmen are points of contact between buyers and suppliers. Thus, a supplier needs to have the right person for this position (Gorchels, 2004, p. 163).

3.4.4 Supplier's Profile

The history and reputation of the suppliers have a great impact on its selection. Buyers will look at a supplier's profile and compare it to other suppliers. Some of the important characteristics of suppliers are:

Reputation: The response of the customers towards the supplier is one of the important factors to decide the performance of the supplier. Suppliers with a good customer base are preferred over the others with no satisfactory customers (Gorchels, 2004, p. 123).

Performance history: The performance history of the supplier also influences its selection. Firms pay attention to the competitive nature of the supplier, its past production schedule, response to market, and its ability to make commercial relations and business references (Gorchels, 2004, p. 122).

Production facility and capacity: The purchasing firms would also take into account the production facilities and ability of a supplier to increase its capacity and compare it with other suppliers. The best would be chosen based on its capacity that matched the purchasing firm's demand (Gorchels, 2004, p.122).

3.4.5 Bringing It Together

Supply chain management is the umbrella that covers all aspects of the sourcing and procurement of goods. Supply chain management forms and manages the business-to-business links that allow for the ultimate sale of goods to consumers.

Throughout the history of the development of supply chain management studies, one could observe three major movements which included: Creation, Integration, and Globalization. In the Creation Era, firms went through the process of re-engineering and downsizing in an attempt to reduce costs. In Integration Era, firms started to use systems such as Electronic Data Interchange (EDI) and Enterprise Resource Planning (ERP) to increase value-adding and cost reductions through integration. Finally in Globalization Era, it can be characterized by the globalization of supply chain management in organizations to increase their competitive advantage, value-adding, and reduce costs through global sourcing.

Given the importance of collaboration between supplier and buyer in the supply chain, firms now pay close attention to the supplier selecting process. Buyers evaluate their future partners through a set of "supplier selection criteria" before conducting business. The supplier selection criteria could be categorized into two separate categories: Tangible and Intangible. Tangible criteria include price, quality, service performance, human capital, and production facility and capacity. The intangible criteria include reputation, trust, commitment, and performance history.

3.5 Emerging Factors in Supplier Selection Criteria

After reviewing several kinds of literature, we found that price, quality, delivery time remain important; however, communication skills, human capital, information technology, and strategic commitment are also crucial factors in the supplier selecting process.

3.5.1 Communication Skills

From the intangible set of criteria, the communication skills of a supplier in the twenty-first century appear to be the most sought-after characteristic that the purchasing firms look for.

Operational communications in supply chains are often characterized by conflicts. Those conflicts are costly in terms of money, time, and effort (Oosterhuis et al., 2012). Therefore, suppliers with superior communication skills will win the race for supplier selection. Carr and Smeltzer (1999) found evidence of the relationship between effective communication with suppliers and a firm's financial performance.

A supplier must be aware of its partner's strategies to stay competitive in the current uncertain and dynamic environment (Oosterhuis et al., 2012). To do that, they need to communicate frequently and closely with their partner.

Many research studies show that one way of fostering the collaboration between supplier and buyer is to develop activities such as site visits (i.e. supply chain partners are exposed to each other's operations and practices), structured discussions (focused on performance objectives, relationship strategies, and analyses of relevant developments), and the use of advanced communication technologies (e.g. Klimoski & Mohammed, 1994; Rico et al., 2008; Ring & Van de Ven, 1994).

These activities facilitate mutual understanding among partners and help to develop better recognition of partners'context and strategies. Such activities are essential during the initial phase of a supply chain relationship (Ring & Van de Ven, 1994), but they are also useful when the relationship matures and the context and performance objectives evolve.

3.5.2 Human Capital

Another important factor that suppliers nowadays must pay close attention to is human capital. In a supply chain relationship, many individual employees are involved. They are salespeople, purchasers, material planners, product developers, logistics managers, sales managers, and directors (Oosterhuis et al., 2012). All of them all need to communicate to some extent with the buyers. A contact person's experience in working with another contact person is a valuable asset in enhancing the partnership. As a result, a supplier may want to ensure a certain degree of stability by retaining the same contact persons over time (Bendapudi & Leone, 2002; Rico et al., 2008) because the turnover of contact persons may lead to a loss of shared knowledge and may hinder goal.

If employee turnover cannot be avoided, the supplier should ensure that they retain the knowledge that contact persons possess (Bendapudi & Leone, 2002). One way of doing so is to motivate employees to share information with supervisors and colleagues or use technology such as ERP systems to record relevant information and create organizational structures that support information sharing. These steps could ensure that suppliers can maintain shared knowledge and perceptions even if contact persons can no longer be retained (Bendapudi & Leone, 2002; Sambasivan et al., 2011).

3.5.3 Information Technology (IT)

Due to globalization, information management is assuming key importance in supply chain management strategy. As a result, the use of IT should be effectively used among all supply chain partners to avoid that poor IT resource management by one or more partners in the supply chain. That poor IT resource management could have negative repercussions on the performance of the entire supply chain in terms of planning ability, costs, and customer service (Ovalle & Marquez, 2003).

Suppliers are currently required to manage information flows along the entire supply chain. Thus, suppliers need investment in IT applications. According to Evangelista (p. 177), there is a positive correlation between the adoption of data gathering and enterprise information technologies and suppliers' performance. In summary, information technology investment is a critical area to achieve competitive advantage.

3.5.4 Strategic Commitment

According to Kannan (1998), another key factor buyers look for nowadays is the strategic commitment from a supplier because it is an important determinant of business success. Strategic commitment not only directly enhances performance but also has an indirect impact. That is, if there is a relationship between buyer and supplier, and if there are shared expectations and objectives then it is easier to address supplier delivery and quality problems. Buyers who search for long-term suppliers are looking for this characteristic because they require suppliers that are willing to develop closer ties, have order entry systems that support the relationship, are willing to share confidential information, and are committed to serving their long-term needs. By doing so, both parties are benefitted. Tracey and Tan (2001) found that the involvement of suppliers in the buyer's product development process and continuous improvement programs increase customer satisfaction and the overall firm performance. In Fig. 3.1, we present our conceptual model that links the traditional and emerging factors for selecting the suppliers.

3.6 Supplier Selection in Automotive Industry

As one may notice, the criteria of selecting suppliers depend on the type of product and industry. Often the process involves a trade-off between the different dimensions such as quality and cost. In this section, the paper will narrow down the topic and review several pieces of literature to find out: "*What are the characteristics of a top-tier automotive supplier?*"

The automotive industry in North America has gone through significant changes in the past decades. As stated in a study by Brunnermeie and Martin (2002), in response to Japanese competition, US automakers are reducing the time and cost of new product development by adopting the philosophies of core competence and concurrent engineering. They are increasingly focusing on parts and services in which they possess a clear competitive advantage and are outsourcing other work. Even though there are over 15,000 components in an average automobile, only a few are manufactured by the final assemblers, the majority being supplied by a network

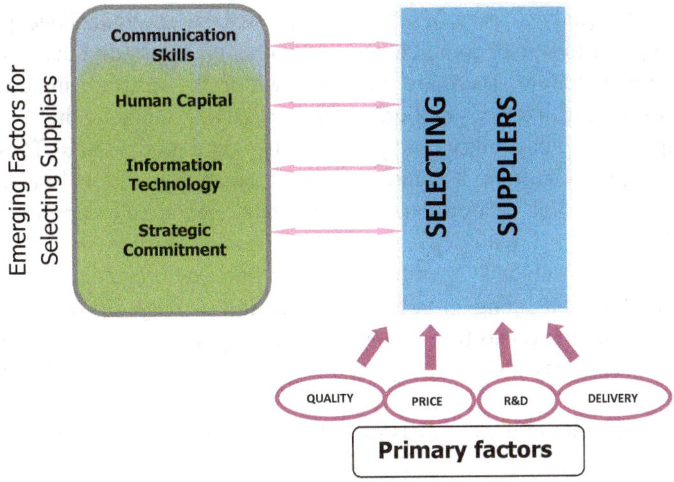

Fig. 3.1 The conceptual model for supplier selection

of specialist component manufacturers (Manuela & Angel, 2001). The adoption of these philosophies has forced significant changes in the relationships between the firms and their suppliers. Many firms now require that their suppliers support their preferred systems so that transfers can be made to and from the supplier in native format.

In addition, car manufacturers are likely to demand that suppliers assume substantial responsibility during product development; accommodate customer requests for engineering changes in their product or manufacturing process; be highly reliable concerning quality and delivery, and have the ability to respond quickly in case of problems. These requirements are difficult to meet unless the suppliers have adopted lean production themselves. Thus, a lean customer is likely to find it more productive to work with lean suppliers, and therefore to diffuse lean production through the supply chain (MacDuffie & Helper, 1997).

Before the emergence of lean production techniques, buyer-supplier relationships featured short-term contracts, a high number of suppliers for each component, and competition among suppliers based almost exclusively on price. Nevertheless, current standards in the automotive industry show that many more contracts are extended throughout the entire vehicle life-cycle, that number of suppliers by components has been reduced, and that the competition is based primarily on quality, engineer capability, and just in time delivery (Manuela & Angel, 2001).

Just in time (JIT) is a strategic part of the automotive industry, and JIT procurement and production systems continue to grow in number and importance. Bartholomew reported as early as 1984 that "all of the major US automotive firms profess to have an ongoing JIT program at this time".

A study conducted by Doran (2004) investigating the characteristics of the first-tier module suppliers in the automotive industry reveals several important indications.

According to the study, in terms of supply chain activity, the modular concept requires first-tier suppliers to deliver complete modules rather than the individual components that constitute a module. There are three tiers within the industry: mature first-tier supplier, developing first-tier supplier, and fringe first tier. A mature first-tier supplier has the capacity and capability to supply modules on a global basis. They possess a distinctive quality culture, a significant R&D capability, a global presence, and ownership of key areas of a modular supply chain. Developing suppliers are first-tier suppliers (possibly with some second-tier businesses) that are in the process of positioning themselves to supply a modular basis by enhancing their supplier relationship skills, engaging in supplier acquisition activity. Fringe suppliers are likely to be those that could be categorized primarily as second-tier suppliers with some first-tier business and can be considered as marginal players within a modular context.

The characteristics of the suppliers in each tier can be summarized in the following Table 3.1.

Doran (2004) also suggested a valuable implication that those suppliers that fail to recognize the importance of rethinking their roles within a modular supply chain are likely to be subsumed as other suppliers seek to enhance their module offerings.

3.6.1 Characteristics of a First Tier Supplier in the Automotive Industry

In summary, from the literature reviewed above, to be a first-tier supplier in the automotive industry, one must have the following characteristics:

- High level of technology
- Involve in product development
- Reliable to quality
- Just in time delivery
- Long-term supplier
- Significant R&D resource
- Brand name
- Global presence

In developing and sustaining such competitive characteristics for a firm in the dynamic automotive industry, the role of emerging factors as listed in our conceptual model: communication skills, human capital, information technology, and strategic commitment is critical. In the following section, we illustrate how these emerging factors are shaping the winning formula for firms in the automotive industry, a very competitive and dynamic industry, especially in the face of the COVID-19 disruptive event.

Table 3.1 Continuum of first-tier suppliers (Doran, 2004)

Mature first tier	Developing the first tier	Fringe first tier
Key resources: • High levels of technology • A global presence • Well-developed knowledge management capabilities • A distinctive R&D orientation • The ability to become a modular supplier and maybe a "module seller" • A branded product or product range	Key resources: High levels of technology Medium-sized operations with some overseas operations Developing knowledge management capability Developing brand name Significant R&D resource	Key resources: Low levels of technology Small scale operation with no global presence Poor knowledge management capability No brand identity
Supplier relationship capabilities: May control key elements of the supply chain, particular where the supplier is developing its position as a modular supplier	Supplier relationship capabilities: Developing supplier relationships Represented in many of the key production regions but may have limited global supply chain presence	Supplier relationship capabilities: Limited to directives issued by its OEM customers Poorly positioned to accommodate the changes necessary
Strategic positioning: Has a clear focus upon key changes within the sector and has commenced with investments necessary to position itself as a major player within a modular environment	Strategic positioning: A proactive player that is aware of key changes within the sector and is positioning itself to provide solutions commensurate with such change	Strategic Positioning: A reactive player with no long term view of changes taking place within the sector
Inter-organization relations: Enjoys contractual trust with its OEM customers Provides quality assured parts/systems on a JIT or synchronous basis	Inter-organizational relations: Enjoys contractual trust with its OEM customers Provides quality assured parts on a JIT basic	Inter-organizational relations: OEMs dictate terms and conditions and impose penalties for non-compliance Parts subject to OEM inspection
Operations: Single sector focus Flexible operations with the ability to "direct line" feed Stage three to four of the Hayes and Wheelright model Order winners likely to include global reach, well-developed R&D expertise, branded product range	Operations: Single sector focus Flexible operations Stage three of the Hayes and Wheelwright model Order winners likely to include: global reach, and the ability to provide modular solutions on a synchronous basis	Operations: Maybe multi-sectoral Non-flexible operations Positioned at stage one or two of the Hayes and Wheelwright model No real order-winners
Value transfer potential Likely to be high	Value transfer potential: Likely to be high	Value transfer potential: Limited since value creation is low

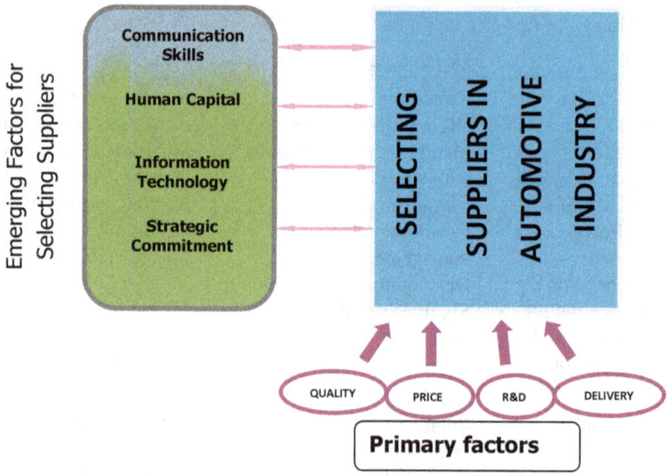

Fig. 3.2 The application of model in supplier selection in automotive industry

3.6.2 Application of the Supplier Selection Model to Automotive Industry

Here we apply our model's emerging factors to analyze the case of the automotive industry, as shown in Fig. 3.2.

3.6.3 Impact of Communication Skills on the Automotive Industry

Communication skills play a major role in determining the success of a company in the automotive industry. Matson and Matson (2007) found that the third most frequently cited problem in Just in Time implementation was the supplier's inability to deliver materials on time. According to Yasin et al. (2007), the cause of this problem is a lack of communication between the company and its suppliers.

Monczka et al. (1998) investigated the importance of communication behavior, such as information quality for the success of supplier alliance. Their study showed that accuracy, timeliness and adequacy, and credibility of information have a positive impact on supplier alliance success aspects such as satisfaction, price, quality, cycle time, and technology.

Rossin (2007) explored the critical role of information quality for the success of efficient and responsive supply chains. He identified that, amongst other factors, poor information quality results in increased inventory, an increase in total costs, and a degrading of customer service due to missing and delayed order product development time.

According to Wiengarten et al. (2010) unless the exchanged information is of high quality they cannot expect a high return from their collaborative initiatives in terms of improved operational performance. The quality of information is dependent on its timeliness, accuracy, relevance, and added value. Making joint decisions and sharing risks and benefits throughout the supply chain does not improve a company's operational performance if the exchanged information is of poor quality

For companies surveyed across the tiers of the supply chain in Wiengarten et al. (2010) first model using low information quality collaborative practices only explained 15% of the variance in operational performance. However, in Weingarten et al.'s (2010) the second model using the sample of companies exchanging high-quality information collaborative practices explained 25% of the variance in operational performance.

In summary, communication influences performance: it shortens development time (Dyer, 1996), enhances part design quality (Takeishi, 2001), reduces uncertainty (Daft & Lengel, 1986), and improves development performance and delivery schedule compliance (Brown & Eisenhardt, 1995)—effectively contributing to the minimizing of COVID-19 like detrimental disruptions of supply chains. Therefore, effective communication can be a source of competitive advantage (Li et al., 2006). Suppliers have to develop and strengthen their communication technologies and skills to win businesses across the globe.

3.6.4 Impact of Human Capital on Automotive Industry

Human capital is an important asset in the automotive industry. For effective communication, suppliers need employees with effective communication skills. In addition, the requirement of JIT delivery has made many suppliers adopt the lean production techniques and as a result in a lean environment, a capable workforce with multiskilled workers is needed to handle the increased responsibility that JIT delivery requires and to push for continuous improvement (Manuela & Angel, 2001).

According to Manuela and Angel (2001), operation managers emphasized training as a just-in-time requirement because it made it easier to solve problems and improve the production process to comply with this strict delivery system. Training and employee empowerment are a critical component that contributes to the success of lean production. Therefore, suppliers need to invest in their human capital to be successful in the industry.

3.6.5 Impact of Information Technology on the Automotive Industry

Information technology plays a significant role in determining the success of a first-tier supplier in the automotive industry. According to Brunnermeie and Martin (2002) the responsibility for the design of an automobile and the factory that produces it is now distributed among many companies; thus, product data must be shared among a greater number of people and organizations both concurrently and sequentially. One auto original equipment manufacturer (OEM) estimates that as many as 453,000 product data exchanges (PDEs) occur each year within the company and its suppliers. Another OEM estimates that electronic exchange of computer aid design (CAD) data alone occurs at least 7000 times per month; that quantity rises as high as 16,000 transfers per month during peaks. This study also estimates that imperfect interoperability costs the US automotive industry approximately \$1 billion per year and delays the introduction of new models by at least two months. While these costs comprise less than one-tenth of 1% of revenues for both the OEMs and the first-tier suppliers, they represent a much higher cost burden for the tooling suppliers.

In addition, suppliers have to meet the requirements of their customers. Uneven and unpredictable demands may cause suppliers to hold higher levels of inventory. As a result, an effective and efficient information sharing system is needed in place.

On the same side of the coin, nowadays, more automotive suppliers transfer information to their customers more frequently (Helper & Sako, 1995). Suppliers were sharing information with their customers: description of their production process, production planning, and quality control methods used, and the costs structure of each production stage. It has become the norm of the industry. The information on costs helped to establish long-term contracts because the auto manufacturer knew how much their suppliers could reduce prices without losses and when and how to help them with performance programs (Manuela & Angel, 2001). The process of knowledge accumulation improves the competitive performance of organizations in the auto industry (Vekstein, 1998).

Yun (1999) found in the Korean automotive industry that prime contractors pick those suppliers with relatively greater technical competence and potential. In the U.S. automobile industry, there are strong competitive pressures such that delays in the transit of information and goods should be eliminated in the industry's JIT environment. Thus, since the mid-1990s, the "Big Three" have been strongly encouraging their suppliers to use EDI (Kurokawa et al., 2008). For example, GM has a substantial track record in supplying complete automation systems by using EDI to its vendors. The company runs a global network called EDSNET, linking more than 30 GM data processing centers with over 2000 suppliers via EDI (Kurokawa et al., 2008). In Another example, Ford launched the Ford Supplier Network (FSN) in 1998, which consists of 80 custom applications supporting more than 4200 suppliers and has approximately 42,000 end users globally (Kurokawa et al., 2008). It is said that Toyota completed all the transactions with its first-tier suppliers by EDI in 2001

(Kurokawa et al., 2008). In summary, information technology has had a great impact on the supply chains of the automotive industry, especially in this COVID-19 era.

3.6.6 Impact of Strategic Commitment on Automotive Industry

As mentioned above in supplier selection in the automotive industry, auto manufacturers are looking for long-term suppliers. As a result, a strategic commitment is a must. Some automotive manufacturers have established programs so that the suppliers may contribute with their ideas to reduce costs or improve the quality. The supplier may share the savings or the profits that the automaker obtains from the idea that will improve its assessment and thus assure it a greater chance of gaining future business. The manufacturers can therefore achieve economies thanks to the innovation inputs from their suppliers (Frey & Schlosser, 1993).

In the USA and Japan, between 25 and 50% of the suppliers surveyed are involved in component design (Cusumano & Takeishi, 1991). The principal advantage that supplier involvement in component design has for an automotive manufacturer is that it helps to reduce the new model development time and costs, and improve product design and design for manufacturability.

For the suppliers, there are also advantages because they have more time to overcome any problems faced in the design-to-manufacturing transition. Innovation is yet another benefit to supplier involvement. According to Manuela and Angel (2001), one of the two involved suppliers in the Spanish automotive industry had developed and produced components under tight specifications for more than ten years. Under the new regime, it could now access the central computer aid design (CAD) data files and inspect the broader environment surrounding its subsystem. As a result, the supplier developed, tested, and submitted a new design that could save 10% in costs.

Therefore, strategic commitment is a key to success in supply chain collaboration. When companies could not get suppliers to cooperate; they changed suppliers (Matson & Matson, 2007). In the Post-COVID era, this collaborative capability has become more relevant and needed for the business organization to deal with the risks and uncertainties of such external shocks.

3.7 Implication for Supplier Selection in the Post-COVID Era

Critical analysis and review of literature on suppliers' selections informed the construction of our dynamic model that links the traditional and emerging factors. Ambidextrous learning and application of this model are shown to improve the performance supply chains of the automotive industry. When we look at the damage and disruptions that COVID-19 has inflicted on businesses including on the automotive industry, the role of emerging factors including (Kiers et al., 2022) in suppliers selection have become even more critical:

- **Communication Skills**: Suppliers with superior communication technologies and capabilities should be able to manage and reduce supply chain risks about market shocks like COVID-19. Capabilities development strategy should be linked to human capital and strategic commitment initiatives. Communication technologies should be in sync with the firm's IT strategy.
- **Human Capital**: Develop human capital with diverse, complementary skills and competencies levels. With an increased level of competencies, the workforce can improve supply chain resilience- a much-needed capability to withstand and deal with COVID-19 like disruptions. A supply chain's coordination capability is heavily dependent on the human capital of the suppliers (and their client organizations).
- **Information Technology**: To overcome the challenges of uncertain disruptions like COVID-19, the deployment of information technology (IT) play a critical role in the overall performance of a supply chain. IT plays a promising and critical role in developing a firm's communications skills and human capital. Suppliers with top-line IT technologies are more likely to win the businesses of their client firms.
- **Strategic Commitment**: When collaboration is proved as a winning strategy for suppliers, the role of stratic commitment can not be underscored. Resource pooling and risk sharing among supply chain partners including suppliers in a Post-COVID-19 era are likely to be the most needed capability. Ensuring solid strategic commitment on the part of suppliers and their clients is critical for the management of risks associated with disruptions like COVID-19.

Overall, to better manage the risks and uncertainties in a Post-COVID world, suppliers' selection appears to be dependent on how well they score on these four critical, dynamic, and integrated factors: communication skills, human capital, information technology, and strategic comment.

3.8 Conclusion

In this fast-moving world, managing an effective supply chain is a complex task for many companies. It requires not only time, money, effort but also collaboration from both their upstream suppliers and downstream customers. Thus, firms now pay close attention to the process of choosing their partners.

The purpose of this paper is to review the supplier selection process of purchasing firms and from that, determine the recent trends in the characteristics of a long-term supplier that companies are searching for. According to the research, buyer firms have different sets of criteria for the supplier selection process. The criteria could be categorized into two separate categories: Tangible and Intangible. Tangible criteria include factors such as price, quality, delivery time… while intangible criteria include factors such as supplier's reputation, communication skills, commitment… Both sets represent the key characteristics that purchasing firms are looking for in a supplier.

While cost, quality, and delivery performance have been consistently identified as being important determinants of supplier selection, communication skills, information technology, human capital, and strategic commitment increasingly demonstrate their importance in the supplier selection process. The complex nature of supply chain management requires effective and efficient communications from both parties; thus, companies are searching for suppliers with great communication skills. Information technology and human capital also became the key characteristics for effective communication. The need for effective communication requires suppliers to invest in their information technology systems to better control the flow of information between the parties. The staffs who directly deal with the buyers possess shared knowledge of both supplier and buyer. Having a stable workforce ensures that knowledge is retained. Last but not least, firms now require more collaboration from their suppliers; thus, strategic commitment has become crucial.

In the automotive industry, key characteristics and capabilities include a high level of technology, involvement in product development, reliability to quality, just in time delivery, long-term supplier, significant R&D resource, brand name, and global presence. However, emerging factors (communication skills, information technology, human capital, and strategic commitment) also apply to the industry. Automakers are selecting suppliers with those skills. In uncertain times like now and Post-COVID-era, all partners in any supply chain will be well served by making solid investments and efforts to develop these unique but integrated skills, capabilities, and resources.

Acknowledgements Author would like acknowledge and thank his student, Mr. Tien Li for his assistance in the earlier draft of this chapter.

Appendix 1

Survey Items

How important are the following factors when selecting a key/preferred supplier for your organization?

		High				Low
a.	Company size	5	4	3	2	1
b.	Ethical standards	5	4	3	2	1
c.	Testing capability	5	4	3	2	1
d.	Scope of resources	5	4	3	2	1
e.	Technical expertise	5	4	3	2	1

f.	Industry knowledge	5	4	3	2	1
g.	Commitment to quality	5	4	3	2	1
h.	Open to site evaluation	5	4	3	2	1
i.	Supplier's process capability	5	4	3	2	1
j.	Insurance and litigation history	5	4	3	2	1
k.	References/reputation of supplier	5	4	3	2	1
l.	Ability to meet delivery due dates	5	4	3	2	1
m.	Price of materials, parts, and services	5	4	3	2	1
n.	Financial stability and staying power	5	4	3	2	1
o.	Supplier's effort in eliminating waste	5	4	3	2	1
p.	Honest and frequent communications	5	4	3	2	1
q.	Flexible contract terms and conditions	5	4	3	2	1
r.	Geographical compatibility/proximity	5	4	3	2	1
s.	Cultural match between the companies	5	4	3	2	1
t.	Past and current relationship with supplier	5	4	3	2	1
u.	Suppliers' effort in promoting principles JIT	5	4	3	2	1
v.	Supplier has strategic importance to your firm	5	4	3	2	1
w.	Supplier's willingness to share confidential information	5	4	3	2	1
x.	Percentage of supplier's work commonly subcontracted	5	4	3	2	1
y.	Supplier's order entry and invoicing the system, including EDI	5	4	3	2	1
z.	Your annual orders as a percentage of their overall business	5	4	3	2	1
aa.	Supplier's ability to make a decent profit for supplying to you	5	4	3	2	1
bb.	Willingness to integrate supply chain management relationship	5	4	3	2	1
cc.	Commitment to continuous improvement in product and process	5	4	3	2	1
dd.	Reserve capacity or the ability respond to unexpected demand	5	4	3	2	1

How important are the following issues when evaluating your key/preferred suppliers' performance?

		High				Low
a.	Quality level	5	4	3	2	1
b.	Service level	5	4	3	2	1
c.	Correct quantity	5	4	3	2	1
d.	On-time delivery	5	4	3	2	1
e.	Price/cost of product	5	4	3	2	1
f.	Use of electronic data interchange (EDI)	5	4	3	2	1
g.	Willingness to share sensitive information	5	4	3	2	1
h.	Presence of certification or other documentation	5	4	3	2	1
i.	The flexibility to respond to unexpected demand changes	5	4	3	2	1
j.	Communication skills/systems (phone, fax, e-mail, Internet)	5	4	3	2	1
k.	Quick response time in case of emergency, the problem, or special request	5	4	3	2	1
l.	Willingness to change their products and services to meet your changing needs	5	4	3	2	1
m.	Willingness to participate in your firm's new product development and value analysis	5	4	3	2	1

Indicate the level of your firm's performance compared to that of major industrial competitors in terms of:

		High				Low
a.	Market share	5	4	3	2	1
b.	Return on assets	5	4	3	2	1
c.	Overall product quality	5	4	3	2	1
d.	Overall competitive position	5	4	3	2	1

See Tables 3.2 and 3.3.

References

Balachandran, K. (2005). Quality implications of warranties in a supply chain. Retrieved from http://mansci.journal.informs.org/content/51/8/1266.full.pdf [Accessed on February 13, 2022].

Table 3.2 Elements of supplier selection

	Selection criteria	Mean score[a]
l.	Ability to meet delivery due dates	4.62
g.	Commitment to quality	4.60
e.	Technical expertise	4.25
m.	Price of materials, parts, and services	4.16
p.	Honest and frequent communications	4.11
dd.	Reserve capacity or the ability to respond to unexpected demand	4.08
f.	Industry knowledge	4.06
n.	Financial stability and staying power	4.03
i.	Supplier's process capability	3.98
cc.	Commitment to continuous improvement in product and process	3.98
b.	Ethical standards	3.92
h.	Open to site evaluation	3.90
k.	References/reputation of supplier	3.86
q.	Flexible contract terms and conditions	3.79
c.	Testing capability	3.77
v.	Supplier has strategic importance to your firm	3.76
d.	Scope of resources	3.69
t.	Past and current relationship with supplier	3.63
bb.	Willingness to integrate supply chain management relationship	3.39
w.	Supplier's willingness to share confidential information	3.37
o.	Supplier's effort in eliminating waste	3.29
aa.	Supplier's ability to make a decent profit for supplying to you	3.25
u.	Supplier's effort in promoting JIT principles	3.24
z.	Your annual orders as a percentage of their overall business	3.15
j.	Insurance and litigation history	3.14
r.	Geographical compatibility/proximity	3.07
y.	Supplier's order entry and invoicing system, including EDI	3.03
s.	Cultural match between the companies	2.90
x.	Percentage of supplier's work commonly subcontracted	2.87
a.	Company size	2.67

[a]Lines indicate groupings based on Tukey multiple comparisons

Bendapudi, N., & Leone, R. P. (2002). Managing business-to-business customer relationships following key contact employee turnover in a vendor firm. *Journal of Marketing, 66*(2), 83–101.

Brown, S. L., & Eisenhardt, K. M. (1995). Product development: Past research, present findings, and future directions. *Academy of Management Review, 20*(1), 343–378.

Brunnermeie, S. B., & Martin, S. A. (2002). Interoperability costs in the US automotive supply chain. *Supply Chain Management: An International Journal*, 7(2): 71–82.

Table 3.3 Elements of supplier assessment

	Assessment criteria	Mean score*
a.	Quality level	4.73
b.	Service level	4.62
d.	On-time delivery	4.57
k.	Quick response time in case of emergency, problem, or special request	4.44
i.	The flexibility to respond to unexpected demand changes	4.27
c.	Correct quantity	4.15
e.	Price/cost of product	4.10
l.	Willingness to change their products and services to meet your changing needs	3.88
j.	Communication skills/systems (phone, fax, e-mail, internet)	3.79
m.	Willingness to participate in your firm's new product development and value analysis	3.57
h.	Presence of certification or other documentation	3.50
g.	Willingness to share sensitive information	3.10
f.	Use of electronic data interchange (EDI)	2.69

Chao, G. (2009). Quality improvement incentives and product recall cost-sharing contracts. Retrieved from http://mansci.journal.informs.org/content/55/7/1122.full.pdf [Accessed on February 23, 2022].

Cusumano, M., & Takeishi, A. (1991). Supplier relations and management: A survey of Japanese, Japanese transplant, and US auto plants. *Strategic Management Journal, 12*(8), 563–588.

Daft, R. L., & Lengel, R. H. (1986). Organizational information requirements, media richness and structural design. *Management Science, 32*(5), 554–571.

Doran, D. (2004). Rethinking the supply chain: An automotive perspective. *Supply Chain Management: an International Journal, 9*(1), 102–109.

Dyer, J. H. (1996). Specialized supplier networks as a source of competitive advantage: Evidence from the auto industry. *Strategic Management Journal, 17*(3), 271–291.

Ellram, L. M. (1990). The supplier selection decision in strategic partnerships. *Journal of Purchasing and Materials Management, 20*(4), 8–14.

Florez-Lopez, R. (2007). Strategic supplier selection in the added-value perspective: A CI approach. *Information Sciences, 177*(5), 1169–1179.

Gorchels, L. (2004). *The manager's guide to distribution channels.* McGraw-Hill.

Helper, S., & Sako, M. (1995). Supplier relations in Japan and the United States: Are they converging. *Sloan Management Review, 36*(3), 77–84.

Hill, A., & Hill, T. (2009). *Manufacturing operations strategy* (3rd ed.). Palgrave Macmillan.

Kaplan, R. (1998). *Cost and effect: Using integrated cost system to drive profitability and performance.* Harvard Business School Press. p.13.

Kiers, J., Seinhorst, J., Zwanenburg, M., & Stek, K. (2022). Which strategies and corresponding competencies are needed to improve supply chain resilience: A COVID-19 based review. *Logistics, 6*(12), 1–17.

Klimoski, R., & Mohammed, S. (1994). Team mental model—Construct or metaphor. *Journal of Management, 20*(2), 403–437.

Kurokawa, S., Manabe, S., & Rassameethes, B. (2008). Determinants of EDI adoption and integration by US and Japanese automobile suppliers. *Journal of Organizational Computing and Electronic Commerce, 18*(1), 1–33.

Lam, K.-C., Tao, R., & Lam, M.C.-K. (2010). A material supplier selection model for property developers using fuzzy principal component analysis. *Automation in Construction, 19*(5), 608–618.

Lee, M. (2003). A high-quality-supplier selection model for supply chain management and ISO 9001. *Systems, 14*(3), 225–232.

Lee, E. K., Ha, S., & Kim, S. K. (2001). Supplier selection and management system considering relationships in supply chain management. *IEEE transactions on Engineering Management*, *48*(3), 307–318. https://doi.org/10.1109/17.946529

Li, S., Ragu-Nathan, B., Ragu-Nathan, T. S., & Subba Rao, S. (2006). The impact of supply chain management practices on competitive advantage and organizational performance. *Omega, 34*(2), 107–124.

Lijuan, M. (2002). The primary study of supplier selection based on supply chain management. *Industrial Engineering Management, 6*, 23–25.

Lummus, R. R., & Vokurka, R. J. (1999). Defining supply chain management: A historical perspective and practical guidelines. *Industrial Management & Data Systems, 99*(1), 11–17.

MacDuffie, P., & Helper, S. (1997). Creating lean suppliers: Diffusing lean production through the supply chain. *California Management Review, 39*(4), 118–151.

Manuela, P., & Angel, M. (2001) Internet, lean production and supply chain management in the automotive supplier industry. *Supply Chain Forum: An International Journal, 2* (2), 2–11.

Matson, J. E., & Matson, J. O. (2007). Just-in-time implementation issues among automotive suppliers in the southern USA. *Supply Chain Management: An International Journal, 12*(6), 432–443.

Mentzer, J. T., et al. (2001). Defining supply chain management. *Journal of Business Logistics, 22*(2), 1–25.

Monczka, R. M., Peterson, K. J., Handfield, R. B., & Ragatz, G. L. (1998). Success factors in strategic supplier alliances: The buying company perspective. *Decision Sciences, 29*(1), 69–73.

Movahedi, B., Lavassani, K., & Kumar, V. (2009). Transition to B2B e-marketplace enabled supply chain: Readiness assessment and success factors. *The International Journal of Technology, Knowledge and Society, 5*(3), 75–88.

Oosterhuis, M., Vaart, T. V. D., & Molleman, E. (2012). The value of upstream recognition of goals in supply chains. *Supply Chain Management: An International Journal, 17*(6), 582–595.

Ovalle, O. R., & Marquez, A. C. (2003). The effectiveness of using e-collaboration tools in the supply chain: An assessment study with systems dynamics. *Journal of Purchasing and Supply Management, 9*(4), 151–163.

Rico, R., Sanchez-Manzanares, M., Gil, F., & Gibson, C. B. (2008). Team coordination processes: A team knowledge-based approach. *Academy of Management Review, 33*(1), 163–185.

Ring, P. S., & Van de Ven, A. H. (1994). Developmental processes of cooperative interorganizational relationships. *Academy of Management Review, 19*(1), 90–118.

Rossin, D. (2007). An exploratory analysis of information quality in supply chains: Efficient and responsive models. *Journal of Global Business Issues, 1*(2), 151–158.

Sajjad, A. (2021). The COVID-19 pandemic, social sustainability, and global supply chain resilience: A review. *Corporate Governance, 21*(2), 1–17.

Sambasivan, M., Siew-Phaik, L., Abidin Mohamed, Z., & Choy Leong, Y. (2011). Impact of interdependence between supply chain partners on strategic alliance outcomes: Role of relational capital as a mediating construct. *Management Decision, 49*(4), 548–569.

Sarkis, J., & Talluri, S. (2002). A model for strategic selection. *Journal of Supply Chain Management, 38*(1), 18–28.

Shihua, M., & Xubin, W. (2002). A method of confirming the weight of attributes for supplier evaluation. *Industrial Engineering Management, 6*, 5–8.

Shuyong, K., & Rongqiu, C. (1998). The relationship of manufacturers and suppliers under JIT. *Journal of Management Engineering, 3*, 46–51.

Takeishi, A. (2001). Bridging inter-and intra-firm boundaries: Management of supplier involvement in automobile product development. *Strategic Management Journal, 21*(5), 403–433.

Terry, H. (2002). *Manufacturing strategy: Text and cases* (3rd ed.). Irwin McGraw-Hill.

Tracey, M., & Tan, C. L. (2001). Empirical analysis of supplier selection and involvement, customer satisfaction, and firm performance. *Supply Chain Management: An International Journal, V6*(4), 174–188.

Ustun, O., & Demirtas, E. A. (2008). Multi-period lot-sizing with supplier selection using achievement scalarizing functions. *Computers, and Industrial Engineering, 54*(4), 918–931.

Vekstein, D. (1998). Managing knowledge and corporate performance: An empirical analysis of the world automobile industry. *Omega International Journal of Management Science, 26*(5), 551–568.

Weber, C., Current, J. R., & Benton, W. C. (1991). Vendor selection criteria and methods. *European Journal of Operational Research, 50*(1), 2–19.

Wiengarten, F., Humphreys, P., Cao, G., Fynes, B., & McKittrick, A. (2010). Collaborative supply chain practices and performance: Exploring the key role of information quality. *Supply Chain Management: An International Journal, 15*(6), 463–473.

Xia, W., & Wu, Z. (2007). Supplier selection with multiple criteria in volume discount environments. *Omega, 35*(5), 494–504.

Yasin, N. M., Noor, M. N., & Mohamad, O. (2007). Does image of country of origin matter to brand equity? *Journal of Product and Brand Management, 16*, 38–48.

Yun, M. (1999). Subcontracting relations in the Korean automotive industry: Risk sharing and technological capability. *International Journal of Industrial Organization, 17*(1), 81–108.

Part III
Understanding Supply Chain Dynamics - Modeling-Based Empirical Solutions

Chapter 4
Sustainable Outcomes Through the Structured Forward Supply Chain: A System Dynamic Approach

Mohammad Shamsuddoha

Abstract The concepts of sustainability and supply chains are critical components for modern businesses as they face enormous competition and manage economic, social, and environmental sustainability. However, the poultry livestock sub-sector has received insufficient attention from academics, according to current literature. As a result, this particular industry is suffering from unstructured supply chain processes, a lack of awareness of the implications of the sustainability concept, and a failure to recycle poultry wastes. Furthermore, the Covid 19 pandemic puts additional strain on this industry and its supply chain. As a result, the current study uses a case study to develop an integrated poultry forward supply chain model. The integration process model is an expanded version derived from real-world scattered processes performed by various supply chain members. With the help of 'system dynamics' and case study method, this quantitative study used the positivist paradigm and 'design science' methodology. The findings revealed that supply chain integration could provide economic and social sustainability and a structured manufacturing process to support the research objectives and questions. At the end of the chapter, the pandemic effects will discuss briefly to determine the future direction.

Keywords Sustainability · Supply chain · Economic activities · Social activities · Methodological framework · Causal model · Stock and flow model · Simulation · Model reliability · Model validity · Employment · Forward supply chain · System dynamics · Poultry · Poultry production · Post-COVID 19 implication · Bangladesh

This chapter is based on the following doctoral thesis: "Integrated supply chain model for sustainable poultry production in Bangladesh: a system dynamics approach," Curtin University, Australia.

M. Shamsuddoha (✉)
Supply Chain Management, Western Illinois University, Macomb, IL, USA
e-mail: m-shamsuddoha@wiu.edu

H. Qudrat-Ullah (ed.), *Understanding the Dynamics of New Normal for Supply Chains*, Understanding Complex Systems,
https://doi.org/10.1007/978-3-031-07333-5_4

4.1 Introduction

Several participants are at various stages of the poultry forward supply chain process from a supply chain perspective (Barratt, 2004). They are, however, uncoordinated and unaware of what to do with their immediate and preceding supply chain members to meet demand requirements optimally. Each member is excessively focused on their processes rather than on the consequences of their ignorance within it. For instance, breeder farmers prioritize increasing day-old chicks production without considering the supply required by immediate distribution houses. Additionally, distribution houses sell their day-old chicks at the lowest profit margin, with no regard for breeder/parent stock concerns about costs, supply, and market volatility.

Bangladesh is a developing country that faces numerous economic, social, and environmental challenges. The poultry industry is a significant livestock subsector in Bangladesh, contributing innumerable benefits to the society and economy (Alam et al., 2009; Asaduzzaman, 2000). Additionally, approximately two billion chickens are produced each year (Rahman, 2012; Saleque, 2013). Regrettably, this industry has been slow to adopt modern supply chain concepts, technologies, and value addition strategies in the procurement of poultry products and waste processing. As a result, this industry misses out on the opportunity for long-term growth. Likewise, the poultry supply chain is dispersed in nature, with each supply chain member responsible for a small fragmented process (Ahuja & Sen, 2007). Due to a lack of coordination among supply chain participants, the cost of production increases.

Additionally, farmers face difficulties estimating market demand, which frequently results in over-and under-production. Without a doubt, excessive and insufficient production create an unsustainable situation in the relevant market. As a result, it is critical to incorporate sustainability and supply chain concepts into the poultry production process. For example, to improve the poultry situation in Bangladesh, the industry must employ dynamic, structured, and integrated supply chain processes and sustainability concepts. Therefore, the researchers designed this study to understand better how sustainable supply chain practices can be used to mitigate socioeconomic challenges.

Given the current state of affairs, the poultry industry must close the gap between the ideal supply chain and current practices. The system dynamics (SD) approach may be appropriate for this research because it aims to develop an interactive and integrative model that ensures maximum coordination among supply chain members. System dynamics can provide a proper extended model along with simulation results for a complex supply chain system (Jain et al., 2009). However, a plethora of research on Bangladesh poultry and the supply chain exists. Additionally, academic writings on the global poultry supply chain are scarce. These existing research gaps prompted the current study to develop an integrated model for achieving sustainable outcomes. This strategy can open new doors for small and medium-sized enterprises (SMEs), thereby increasing employment opportunities and reducing poverty. Socioeconomic factors take precedence in this research: the industry focuses on operating the business to address societal challenges. It is worth noting that developing a sustainable poultry production process is the highest priority, and no one has yet contributed significant procedures in this direction. Such a research gap on supply chains is required to be addressed through integrating various supply chains into a single framework.

4.2 Research Questions and Objectives

To design a sustainable poultry production process, this study develops an integrated supply chain model. On behalf of this study, we have established the following three research objectives.

RO1: To design a sustainable poultry production process based on existing supply chain processes that are currently dispersed.

RO2: To create a forward supply chain model in the simulation platform to assess the adaptability of SD processes to real-world conditions.

RO3: To identify economic and social changes that result in job creation and thus poverty reduction through a well-designed forward supply chain (FSC).

4.3 Literature Review

4.3.1 Sustainability

Sustainable development is defined as "meeting the needs at the present without compromising the ability of future generations to meet their own needs" (WCED, 1987, 1). According to the Environmental Protection Agency of the United States (EPA), sustainability defines as preserving or restoring social and ecological capabilities (Elkington, 1994; Jennings & Zandbergen, 1995; Sikdar, 2003). Limited resources (Daily & Ehrlich, 1992; Davis, 1990; Wakeford, 2012); overpopulation (Daily & Ehrlich, 1992; Davis, 1990); poverty (Rhyne, 1998); chaotic industrialization (Barrera-Roldán & Saldvar-Valdés, 2002); dwindling living standards (Munro & Holdgate, 1991); polluted natural resources (Tilman et al., 2002); global climate change (Daily & Ehrlich, 1992; Davis, 1990); global climate change (Tilman et al., 1996) are the main reasons behind considering sustainability. Such problems obstruct companies' profitability by preventing them from reaching their full potential. The "triple bottom line" (social, economic, and environmental, or people, profit, and the planet) is a term used to describe these factors (Elkington, 1994, 2004; Norman & MacDonald, 2004; Peacocka & Shermanb, 2010). To achieve a sustainable process in the poultry industry, triple bottom line concepts must be incorporated into the current production system (Akter & Farrington, 2007, 2008, 2009; Dolberg, 2004; Ward, 2002).

Table 4.1 Economic activities in poultry farming

Indicators: economic factors	Poultry activities
Financial Profitability (Dees, 1998)	Making Optimum Profit
Value Addition (Acs & Armington, 2004; Ahmad & Seymour, 2008; Cobb et al., 2009)	Producing By-products and Chicken Processed Food
Sales and Cost of Goods (Cobb et al., 2009; GRI, 2009)	Maximum Sales and Reducing Cost

4.3.2 Economic Activities in Poultry Farming

The primary source of concern in achieving sustainability is economic gain. Therefore, the company is constantly working to fulfill other social and environmental responsibility responsibilities to reach its maximum targeted profit. Three economic activities (Table 4.1) were observed in the poultry industry, which are discussed in the following sections.

4.3.3 Financial Profitability

The primary motivation for operating a business is profit, even if it is a social enterprise dedicated to poverty eradication (Dees, 1998; Seelos & Mair, 2007). Without profit, no business can be sustained to accomplish its short- and long-term goals: the poultry industry is no exception to this rule. The case industry's primary product is day-old chicks. Thus, the current market price of chicks dictates their ultimate profit or loss. Although the day-old-chicks market is inherently vulnerable, the company can turn a profit at the end of the year.

4.3.4 Value Addition

Value addition entails the provision of additional resources and activities in addition to the primary product to provide additional services to customers (Cobb et al., 2009; Grönroos, 1997). Along with the primary product of day-old chicks, this poultry process generates by-products. Poultry meat processing units manufacture a wide variety of value-added products such as chicken nuggets, samosas, and others that fall outside the scope of the research. The input–output poultry model from the literature illustrates value-added chicken foods and by-products from wastes to illustrate the extent to which these can be processed.

4.3.5 Sales and Cost of Goods

Sales and cost of goods are important indicators for determining whether or not a company is sustainable (Cobb et al., 2009; GRI, 2009). Sales are the lifeblood of any business, and the cost of goods significantly impacts sales and profit. For example, the case industry produces a certain number of chicks that need to be sold in the market; otherwise, they will be unable to reduce costs and maximize profits. At the same time, cutting costs is a top priority for the company because it allows it to make more money. By utilizing indigenous raw materials from its sources, the case industry has taken numerous steps to reduce its cost of goods. Biogas, for example, is used to generate heat for brooding the chicks, and bamboo slats are used instead of expensive imported plastic slats. This is how the poultry industry saves money to increase profits to fund all necessary expansions and incentives for its workers and society.

4.4 Social Activities in Poultry Farming

One of the components of sustainability is social benefits, which are primarily a concern for society and the community. In GRI, AICHE, and Dow Jones reports, social indicators in sustainability concepts are listed. A sustainable poultry supply chain process can provide several direct and indirect social benefits. Table 4.2 depicts the social indicators that correspond to the real-world poultry supply chain. The benefits of poultry processing on a direct social level are discussed below.

Table 4.2 Social activities in poultry farming

Indicators: social factors	Poultry activities
Employment Creation (Cobb et al., 2009; GRI, 2009; Knoepfel, 2001; Reynolds et al., 2001)	Parent Stock Farming, Hatchery, Middlemen, Ultimate (Broiler) Farming
Poverty Reduction (Coulthard et al., 2011; Krantz, 2001; Rhyne, 1998; Yunus, 2007)	Farming, Working, Participating as Supply Chain Member
Create Self-employed Young Entrepreneurs (Åstebro & Thompson, 2011; Cobb et al., 2009; Freytag & Thurik, 2010; GRI, 2009; Knoepfel, 2001; Lazear, 2003; Lynch, 2004; Wagner, 2003)	Farming, Distributor, Agent, Sub-agent, Supplier of Raw Materials, Middlemen (Chicks and Chicken Sellers), By-Products Processor, Ultimate Farmers
Creating New Ventures and Family Business Creation (Dyer & Chu, 2003; Heck & Stafford 2001; Stock & Watson, 2003)	Small and Medium-scale Farming, By-Products Processor, Sub-agent
Social Welfare and Care (Hall et al., 2010; Seelos & Mair 2005; Sundin, 2011)	Gaining Profit, Recycle and Reuse of Poultry Wastes

4.4.1 Employment Creation

One of the primary concerns for research question one is whether the forward supply chain of a poultry process creates jobs. The creation of jobs is a key indicator of a sustainable society's social aspects (Cobb et al., 2009; GRI, 2009; Knoepfel, 2001; Reynolds et al., 2001). An industry's social contribution is proportional to the number of jobs it creates. A job opportunity indicates a person's ability to manage their family. As a result, recognizing an industry for its contributions has a tremendous positive impact on society. In terms of job creation, parent and broiler farming, for example, contribute the most. One person is required to rear 500 parents, whereas one person per 1,000 birds is required in broiler farming. The hatchery unit, as well as the role of the middlemen, generate significant employment. Importantly, as the number of poultry birds raised in a farm unit increases, so will the number of job opportunities.

4.4.2 Poverty Reduction

Poverty reduction is a major concern for the Bangladeshi economy, as more than 40% of the country's population lives in poverty (BBS, 2010; Yunus, 2007). The majority of businesses have no idea how they can help to eradicate poverty (Hancock, 1992; Yunus, 2007). It is not thought necessary for a business to keep track of how it contributes to social issues. However, based on product turnover, people involvement, supply chain networks, and other factors, the government or non-government agencies can easily assess this. The poultry industry makes a significant contribution to both the rural and urban economies. People participated in the case industry supply chain network as workers, agents, and middlemen, among other roles. Workers on participating farms, for example, can escape poverty by performing their jobs to a high standard. Furthermore, a large number of people are employed by supply chain members as workers. The case industry maintains 70% of the country's supply chain networks for distributing day-old chicks. Such a network expands the possibilities for eradicating poverty, particularly for many people living in rural areas.

4.4.3 Create Self-Employed Young Entrepreneurs

Creating self-employed young entrepreneurs is beneficial to society because it gives young people the opportunity to work for themselves: sustainability performance can be measured by the number of entrepreneurs in a business network (Åstebro & Thompson, 2011; Cobb et al., 2009; Freytag & Thurik, 2010; GRI, 2009; Knoepfel, 2001; Lazear, 2003; Lynch, 2004; Wagner, 2003). In Bangladeshi rural and peri-urban areas, there are many young unemployed people. They are looking for suitable employment and small businesses to support themselves. Within the poultry supply chain network, there is a good chance of attracting young entrepreneurs. Because of the low capital and space requirements, ultimate (broiler) farming, for example, is a good fit for young entrepreneurs. According to the poultry owner, one young entrepreneur can start rearing 1,000 day-old chicks for around A$3,000, which is a fraction of the cost of other businesses. It is expected that if they can efficiently care for 1,000 birds, they will be able to support themselves and their family. To maintain their self-employed small poultry farming business, they must consider various types of calamities, disasters, and market fluctuations. Distributors, agents, sub-agents, suppliers (raw materials), middlemen (chicks and chicken sellers), and by-products processors all have similar opportunities for self-employment.

4.4.4 Creating New Ventures and Family Business Creation

Creating new ventures is critical for a society because it increases the opportunities for more people and entrepreneurs to participate (Dyer & Chu, 2003; Heck & Stafford, 2001; Stock & Watson, 2003). Within a family, new ventures can be formed at the same time. Even the case industry began as a small farmer operating under the family business in 1986. They later expanded their company and turned it into a joint-stock corporation. They are still running their company as a three-person family. Due to their tremendous success over the last two decades, the case industry has become a role model farm for the surrounding community. Family businesses differ from other businesses in that family members can run them with or without the assistance of additional employees. With the help of other family members, one person can run a small farm with a capacity of 2,000 birds. Similarly, poultry by-product processing and day-old chick supply sub-agents can be kept within a family's boundaries. With so many opportunities to involve other people from within a society, these initiatives can grow ito new ventures.

4.4.5 Social Welfare and Care

The ultimate responsibility for a company that makes significant profits from an economy is social welfare and care (Hall et al., 2010; Seelos & Mair, 2005; Sundin, 2011). The industry's social benefits are a kind of indirect "payback" to society. For example, the case industry built a primary school and a mosque and provided charity to disadvantaged people in the community.

4.4.6 Sustainability with Economic and Social Benefits

Employment, profit, value addition, and sales are essential aspects of small businesses and are considered sustainable economic performance (Venkataraman, 2002). Some academics have argued that the frequency of employment (turnover) is logically linked to economic performance. Bjerke (2007) found significant links between sales, revenues, and employment to support this viewpoint. Furthermore, many academics have used profit as a measure of economic success (Chen et al., 2007; Davis et al., 2010; Honig, 1998; Kreiser & Davis, 2010). Profit is a blood flow into a business that is required to expand and maintain the current business. Every business prioritizes the economic issue, including the financial performance of its stakeholders. This is true in the poultry industry as well. Integrating all fragmented processes can lead to increased profits by lowering costs and increasing efficiency. On the other hand, sustainability considers security, a modest lifestyle, comfort, and health by maintaining eco-friendly goods and services for social well-being (McMichael et al., 2003). Employment creation, poverty reduction, entrepreneurship development, facilities for young entrepreneurs, and social care (Åstebro & Thompson, 2011; Dyer & Chu, 2003; Reynolds et al., 2001; Yunus, 2007) are some of the ways to achieve these goals (Åstebro & Thompson, 2011; Dyer & Chu, 2003; Reynolds et al., 2001; Yunus, 2007). Finally, social benefits help to close the gap between rich and poor in that society (Butler, 2000; McMichael et al., 2003). According to the GRI, Dow Jones, and IChemE, several social indicators are comparable to the current study context. The associated society can gain above-average social benefits from a sustainable poultry process.

4.4.7 Supply Chain Theory and Models

For the past three decades, the supply chain has been a popular concept in business and academia. The supply chain is a process that begins with the management of raw materials and ends with the final consumer, involving the relevant supplier, retailers, and other parties who provide services to the customer (Cox et al., 1995) as well as various points of consumption (Svensson, 2007). The supply chain is essentially a collection of various parties and processes, such as production and backward and forward processes within a company. The supply chain, once again, includes every step in the process of producing and distributing the final product, from the supplier's supplier to the customer's customer (Cooper et al., 1997; Council, 1999; Ellram & Cooper, 1993; Lummus & Alber, 1997; Lummus & Vokurka, 1999). Furthermore, the supply chain manages the supply–demand situation, raw materials sourcing, the manufacturing system, warehousing and inventory, order management, and distribution to consumers through the process of proper planning, sourcing, manufacture, and delivery (Cooper et al., 1997; Lummus & Vokurka, 1999; Quinn, 1997). As a result, the supply chain is concerned with the entire process, beginning with production and ending with consumption. The poultry processors will benefit from an integrated supply chain by lowering costs and increasing profits.

4.4.8 Poultry Forward Supply Chain (FSC)

Suppliers, carriers, investors, policymakers, and intermediaries all have internal and external partners through FSC. In a nutshell, the forward supply chain (FSC) is the sequential process of transforming raw materials into finished goods (Kocabasoglu et al., 2007). Similarly, the poultry forward supply chain begins with the collection of parent stock breeds, followed by the collection of hatchable eggs from the parent breeder, hatching the eggs in the hatchery, distributing the chicks to farmers via middlemen, rearing them for a set period by farmers, and selling meat and eggs to the ultimate customers. The more stable this supply flow is, the greater the benefits in terms of achieving sustainability. A company's supply chain must be effective and efficient to reach its customers quickly. Structured supply chains help businesses reach customers as well as receive timely product returns and deliveries from suppliers. To increase the degree of integration, receive maximum benefits, create a strategic position, and secure desired profits ahead of their competitors, collaborative and supportive relationships throughout the forward supply chain are required (Fuente et al., 2010). The existing literature on poultry sustainability and supply chain issues is almost non-existent. Even in the livestock domain, a long-term supply chain remains unexplored from a global standpoint. On the other hand, sustainable supply chains have been used in various fields and effectively increase profits. This study attempts to integrate these two concepts into the poultry production process to achieve the greatest economic and social benefits.

4.5 Methodology

This study adheres to positivism, which is associated with the quantitative research method and philosophy (Cresswell, 2003; Johnson & Onwuegbuzie, 2004) that employs data derived from a logical and mathematical stream of valid scientific processes (Fairclough, 2003; Larraâin, 1979). Simultaneously, data and information from empirical evidence from a case industry are gathered to justify and validate the ongoing research (Cohen et al., 2011). As a result, the positivist paradigm is concerned with the quantitative research used in this study. To generate analyzes of the proposed supply chain process model, this study used system dynamics (SD) under quantitative methodology with the help of a simulation tool. In-depth interviews, focus group discussions, and observation techniques were used to identify the key variables. Historical data were examined to determine the past trend and behavior for a specific(s) variable, such as the production of poultry chicks weekly.

Table 4.3 Methodological framework (Sterman, 2000; Wolstenholme, 1990)

Phases	Steps	Things to do
Phase 1	Problem Definition and structure	Behaviour-over-time graph development
Phase 2	Identify key variables	Identify key variables through in-depth interview approach
Phase 3	Build causal loop model (qualitative)	Connect identified variables based on cause and effect relationships. And build a complete causal loop model with positive and negative loops
Phase 4	Stock and flow (quantitative) model	Stock (level) and flow (rate) are added to the model variables to build a workable model
Phase 5	Run simulation	Enter real/field data with starting variable
Phase 6	Model reliability and validity (Barlas, 1996; Sterman, 2000, 2001)	Examine structural validity and assess the data reliability in different phases
Phase 7	Test policy and extreme condition situation (Barlas, 1996; Forrester & Senge, 1980; Sargent, 2005)	Considerable changes of key variable values to observe output reliability
Phase 8	Forecasting future behavior	Model run for 312 weeks and it has 104 weeks of real-world data

The methodological phases used in this study are listed in Table 4.3. The historical performance of key variables is first examined. For example, production and supply forwarding to another supply-chain member has shown too much oscillation in the past. At the same time, key variables based on real-world experience with the supply chain process are listed. Following that, a causal loop is created by observing the practical relationship between the variables. Stocks and flow rates are immediately determined to construct a complete simulation (stock and flow) model. The most difficult task is to run the model with data that closely resembles real-world data. To validate the research, reliability, validity, and policy experiments are discussed. Furthermore, appropriate research tools are required to run the simulated model to observe the behavior.

4.5.1 Causal Model

The poultry supply chain, which begins with the 'pure line' breed and ends with the ultimate consumption of meat and eggs, contains several key variables. Only significant variables that play a significant role in creating dynamic behavior in a poultry process system were considered in this study. The complete view of the poultry causal diagram is shown in Fig. 4.1, where key variables were given as input in order of importance. The variables 'mature parent,' 'parent chicks,' and 'hatchery' have the longest loops in this causal model. In the same causal loop, there are several loops. There are 70 different loops of various lengths under the 'parent

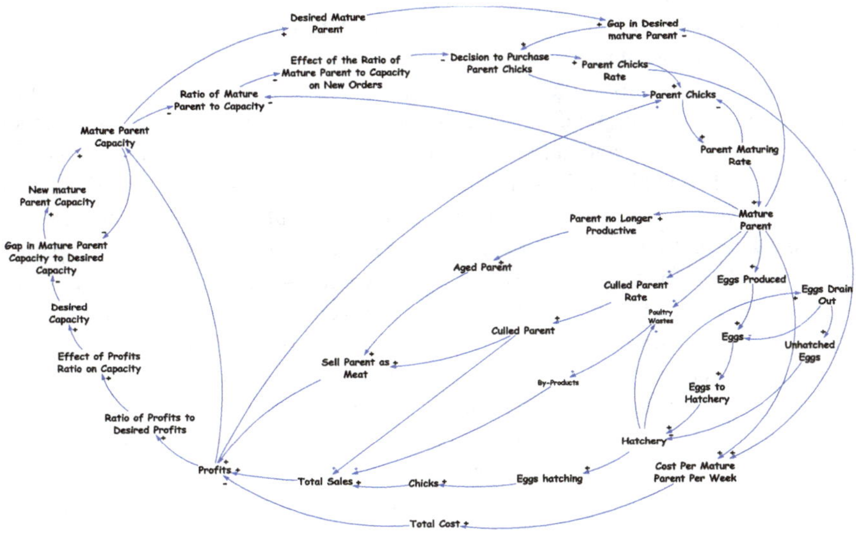

Fig. 4.1 Complete view of poultry causal model

chicks' variable, for example, to express the complex nature of the poultry supply chain. The causal diagram was expressed differently in the above figure, with most of the minor loops identified by adding plus (+) and minus (−) signs.

4.5.2 Stock and Flow Model for Simulation (Quantitative)

It's difficult to say whether the causal loop model accurately reflected reality. Building a model to replicate reality is not the only goal; instead, it is used to experiment until optimality or productive changes are achieved. The quantitative simulation model is the only way to convert the causal loop model into stock and flow. To perceive the model's behavior, a simulation model can add or remove variables. Furthermore, for a good model, validity and reliability are important factors to consider. The model developed and depicted in Fig. 4.2 is the complete stock and flow model for the poultry supply chain within the research boundary. Building a simulation model is a systematic process that takes into account a few key variables. The model is then extended by associating more variables once the major variables interact flawlessly. Even if there is only a minor change in a variable's value, the results of each experiment should be checked as part of this process. Other types of experiments, such as adding or subtracting variables, connecting a new loop to an existing loop, and so on, require a similar procedure. After so much trial and error, this is how perfection is finally achieved. One of the goals of this study was to see if the simulated model performed similarly to the real-life situation in an integrated poultry operation. If the model worked like a real-life poultry operation, it would be used to integrate the majority of the forward supply chain members for a long-term result.

Different methods for collecting information for variables were used to build an authentic SD model. In-depth interviews and industry records were used to gather data on stocks, inflows, and outflows. Some of the data, such as poultry litter, biogas, fertilizer, and fish feed, is critical to track. In this case, the researcher identified how these items were used for a specific period and then calculated the quantity based on how they were used as raw materials. To gather relevant information for computing the input values for the variables, several strategies were used. A few variables (such as parent mortality, hatching rate, and broken eggs) revealed that the data was highly inconsistent. In this case, the data was reprocessed using a trend graph, and then an average value was selected. If the data series for a particular variable showed inconsistencies and oscillation, a value was computed as a random uniform distribution by computing minimum and maximum values using a noise seed and then input as a random uniform distribution. The connected variables were influenced at random by this random uniform distribution based on the minimum and maximum values provided. The model used random normal distribution on these few occasions, with mean and standard deviations taken into account in addition to minimum and maximum values. Each of the model values was computed in this manner, allowing for a successful simulation run. As a result, all variables were eventually set with individual values, just like in real life. As a result, the simulation model was ready to run and analyze.

4.6 Results

This section of the report discusses the simulation model's results in light of the research goals. The following sections discuss the integrated poultry process as well as key performance variables.

4.6.1 Parent Chicks and Mature Parent

The supreme variables in this research model that belong under the forward supply chain are 'parent chicks' and 'mature parent.' These two objects are responsible for the majority of the outputs and behaviors over time. The more parent chicks that are fed into the system, the higher the expected output. Simultaneously, the rest of the supply chain will be kept busy rearing, distributing, and processing more chicks. In Fig. 4.3, the terms 'parent chicks' and 'mature parent,' as well as their rates, are used to represent simulated behavior over 312 weeks (six years). Numerical numbers on the graph lines indicate individual performances. Eight different graph lines depict their behavior over time, which varies slightly due to seasonal variability, natural disasters, and policy barriers. In addition, system delay was a factor in the

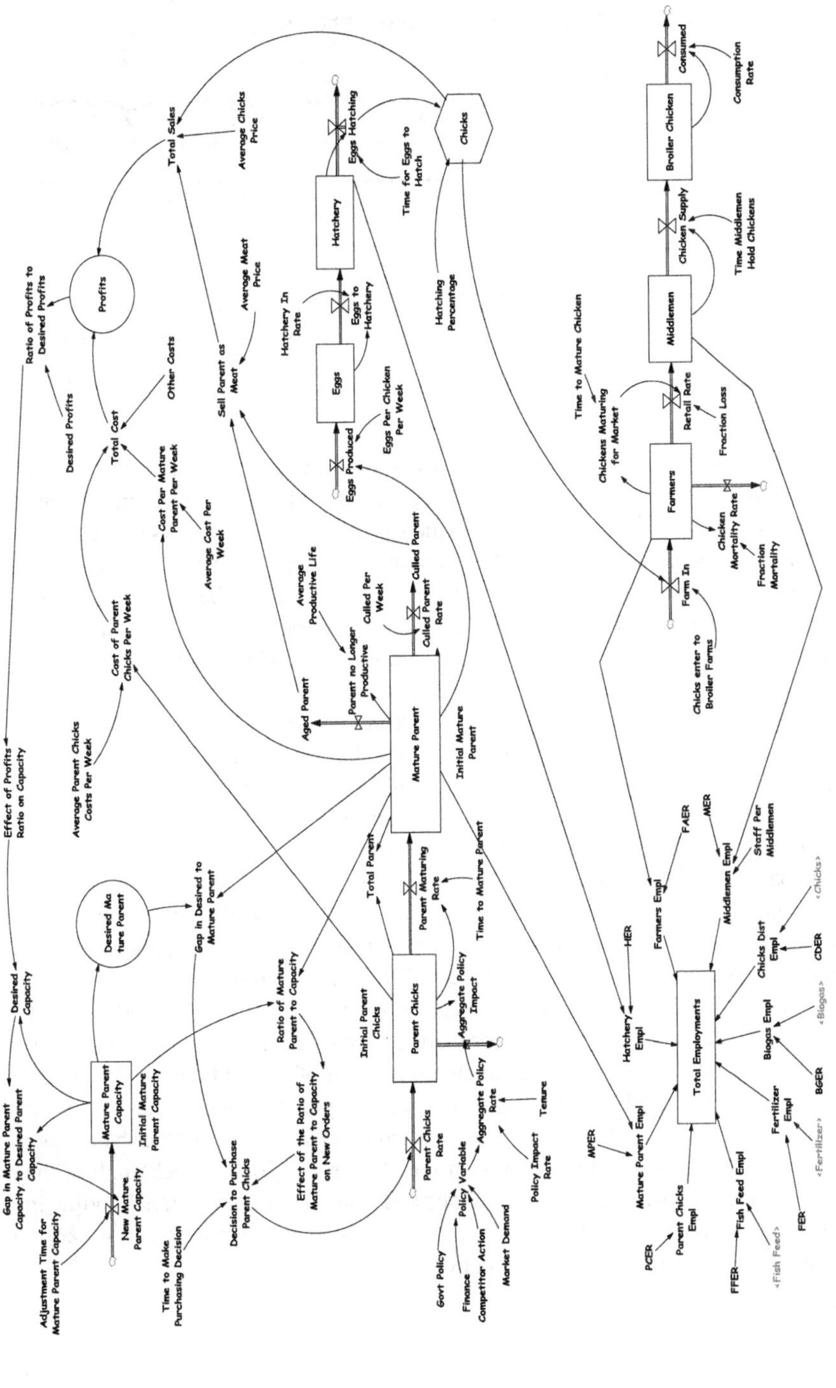

Fig. 4.2 Stock and flow model

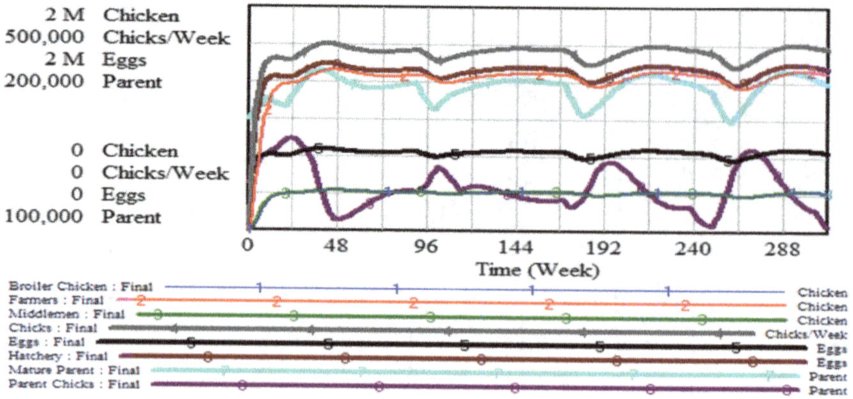

Fig. 4.3 Simulated behaviour of supply chain members

erratic behavior. Notably, in the poultry industry, an uneven variable does not imply an inconsistent operation, as one parent flock stays on the farm for a maximum of 70 weeks, with 40–44 weeks of active production. As a result, 44 weeks of production is roughly equivalent to 85% of a calendar year. If parent chicks are consistently fed into the system, the farm will produce excess chicks that will be unable to sell due to low market demand. This is why farms keep their input based on the parent's life cycle calculation. To maintain their desired amount of production, the case industry usually receives parent chicks two to three times a year.

In Fig. 4.3, lines 7 and 8 represent the mature parent and parent chicks, respectively, while lines 1 and 2 represent broiler chicken and farmers. There are some points on the graph where lines have abruptly risen or fallen, indicating a sudden increase or decrease in demand and production. Maintaining expected demand is difficult because parent chicks grow up in a flock and mature after 24 weeks. The mature parent will continue to produce eggs at different rates depending on their age for another 42 weeks. Between the ages of 30 and 50 weeks, the mature parent produces the most. Figure 4.3 also shows that the mature parent graph line (1) is more consistent than the other graph lines. Parent chicks rate (3) and parent maturing rates (4) are both exhibiting inconsistent behavior due to repeated market adjustments. Maintaining a consistent graph line for mature parents is more important than other graphs because it determines the ultimate production of day-old chicks. The reason for parent chicks' oscillated behavior and mature parent chicks' rate is constant adjustments to the current circumstances. Parent chicks (2), on the other hand, show less fluctuation because they are only in the system for 20–24 weeks before they mature. A flock of parent chicks of various ages can assist the farmers in adjusting to the need. As a result, a parent breeder farm must keep several flocks of varying ages to maintain optimal yearly production.

4.6.2 Day-Old Chicks, Middlemen, Farmers, and Broiler Chicken

Broiler chicken is the ultimate product that has been raised from day-old chicks in a standard environment for a while. Chicks are produced and delivered to farmers through middlemen. As a result, the terms "chicks," "middlemen," "farmers," and "broiler chicken" are all intertwined. Chicks begin to flow to distributors across the country once they have been graded and scientifically packaged. Distributors are stationed throughout the country to ensure that they can reach every corner of Bangladesh. Almost all of the company's major divisions and regions have sales offices. Sub-regional and remote areas are then covered by the sub-agent (also known as a commission agent). The company can maintain a strong supply chain network by supplying day-old chicks to the ultimate farmers in this manner.

When chicks arrive at the ultimate farms, farmers raise them for 4–5 weeks until they reach adulthood. The mature birds are then sold to end-users via middlemen. The model represents these variables' simulated behavior. As the variables are related to the outflow of the previous variable, the graph lines appear identical and follow each other in the end. Thus, the figure is all consistent with the graph lines of chicks, middlemen, farmers, and broiler chickens.

On the other hand, the graph lines are matched, but they produce different numbers of output depending on the requirements. Chick production per week, for example, is around 400,000, while farmers have more chicken in line to sell to the market. Because the birds do not all mature simultaneously, they maintain a consistent weight gain schedule. It all depends on market conditions and whether they will sell mature chickens weighing 1.2 kg or more. If they expect to gain more weight, they may need to give themselves more time to mature. Thus, the model behavior only replicates standard practices and matches the behavior of a real system.

4.6.3 Forward Supply Chain and Employment

One of the main goals of this study is to determine the value of the poultry supply chain to society. The poultry industry's forward supply chain must maintain several operations, including parent farming, hatcheries, feed mills, distributors, broiler farms, and middlemen. At every stage of the process, these operations ostensibly create job opportunities. Figure 4.4 depicts the number of jobs created by each of these operations. Parent farming generates the most jobs, requiring two employees per 1,000 birds, whereas a broiler farm only requires one. The case industry employs more than 150 agents, each with a minimum of four to five employees. Furthermore, the most positive social effects are seen in rural areas, where young unemployed people grow up seeing all of the facilities in the poultry supply chain. As a result, they want to work as middlemen, sub-agents, farmers, feed sellers, and by-products

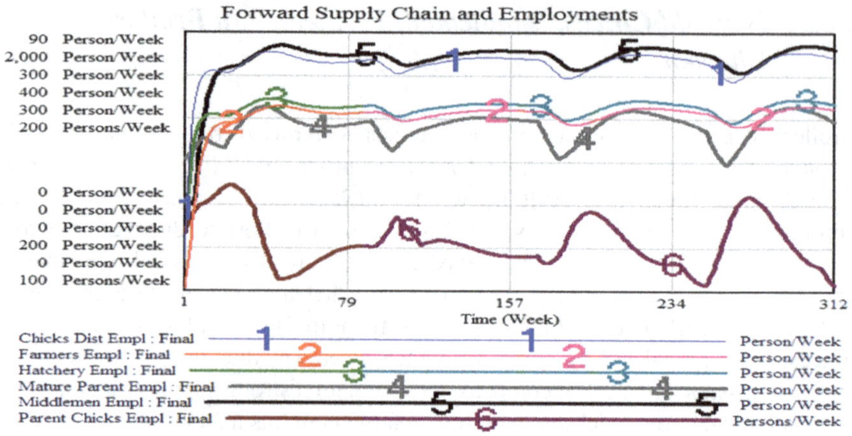

Fig. 4.4 Simulated behaviour for integrated supply chain and employment

processors in the supply chain. This is how the poultry forward supply chain benefits farms, society, and the economy in terms of employment. The opportunities for creating jobs through forwarding supply chain activities are depicted in Fig. 4.4. Among the activities listed are distribution, farming, hatchery unit, parent farm, various middlemen, and breeder (parent) chicks farm. These are the most common activities that can result in job creation based on the number of birds transacted. All of the graph lines follow the behavior of the main activities. The total employment will be calculated based on the behavior of key variables over time. On a farm, 1,000 birds, for example, create one job, while 20,000 birds create 200 jobs.

4.7 Reliability and Validity of the Model

Any type of research, including system dynamics (SD) research, requires a high level of reliability and validity.

4.7.1 Model Reliability

The process of testing policy model reliability is an iterative one (Sterman, 2000). The process of refinement continues until the model meets the requirements for reality and robustness and the ability to reproduce the historical pattern (Forrester & Senge, 1980; Homer & Oliva, 2001; Jørgensen, 2004). The following factors were considered when evaluating the current model's reliability: matching model-generated behavior to real-life data; observing model output by repeatedly changing the noise seed;

conducting multiple runs and comparing results; matching model behavior to real-life trends for different variables. The model passed the above tests and the results were favorable, allowing the model to be declared reliable.

4.7.2 Model Validity

According to Forrester and Senge (1980), Sterman (2000), and Barlas (1996), there is no one-size-fits-all approach to claiming model validity other than comparing real-life and model output, using the extreme condition test, and ensuring the model's dimensional and structural perfection. The majority of SD scholars agree that model validation can only be done by experts who thoroughly understand the real system. Qudrat-Ullah and Seong (2010) have demonstrated six different validation tests based on Forrester and Senge's work once again (1980). These tests can be used to validate the structure of a system dynamics model. The following sections cover boundary adequacy, structure verification, dimensional consistency, and extreme conditions tests to validate the model.

4.7.3 Boundary Adequacy

The model's exogenous variables are thought to be "government policy," "market demand," and "competitor action." The endogenous variables in the model are finance, resource utilization, production, supply chain activities, waste management, demand adjustments, costs, prices and profits, and capital utilization. These variables are appropriate for the industry and the researcher to measure and control. As a result, this model uses an appropriate boundary to deal with a case industry's poultry supply chain.

4.7.4 Structure Verification

The current model was built using data from a real poultry industry, and the model equations and outputs were cross-checked. For example, as the number of parent chicks grows, mature parents grow along with egg production. As a result, the number of parent chicks is decreasing, and they must be replaced with day-old chicks to become mature parents soon. As a result, the causal relationships developed in the supply chain simulation model are validated in a poultry production system using real-world data. Such empirical structure validation (Zebda, 2002) based on real-world knowledge can help the SD model gain confidence.

4.7.5 Dimensional Consistency

Dimensional consistency is a test that determines whether the mathematical equations used in the model variables are dimensionally similar to the existing system (Forrester & Senge, 1980; Qudrat-Ullah & Seong, 2010). The current study checked all of the equations for constant, auxiliary, and level variables to validate the model.

4.7.6 Extreme Condition Tests

The extreme condition (indirect) test is performed by assigning extreme values to selected variables to comprehend model-generated behavior and detect possible real-world behavior under similar conditions (Balci, 1994; Barlas, 1996). To ensure logical behavior in that unusual situation, extreme conditions must be tested using extreme values. A few scenarios were examined below to put the current model to the test in extreme conditions. To understand the model behavior, extreme values were assigned to the associated variables of 'parent chicks.' In reality, the case industry employs about 120,000 new parents and has room for another 200,000 mature parents. When a capacity gap occurs, management is used to taking eight weeks to decide how to fill it. To maintain optimal production, the purchase should include a visionary calculation.

Furthermore, policy barriers, sudden changes in the market situation, the government's uncooperative actions, and other factors stymie around 15% of the business. In an extreme condition, the values of the 'initial parent' and 'mature parent' variables are set to zero (0) and 100,000, respectively, in the 'initial parent' and 'mature parent' variables. In addition, the policy assumed a 30% reduction in business losses, and the decision to purchase new chicks was set for 15 weeks. Figure 4.5 shows the simulated behavior for parent chicks generated by the model run with extreme values. The behavior of this particular variable over time in an extreme situation was revealed by line 1 on the graph. The researchers then discussed the findings with poultry experts and industry executives to ensure that the model's behavior in this extreme condition was accurate.

4.8 Discussions of Results

The main purpose of this research is to integrate a poultry system to get additional economic and social benefits. The below discussions highlight the economic and social benefits of merging fragmented operations under one system.

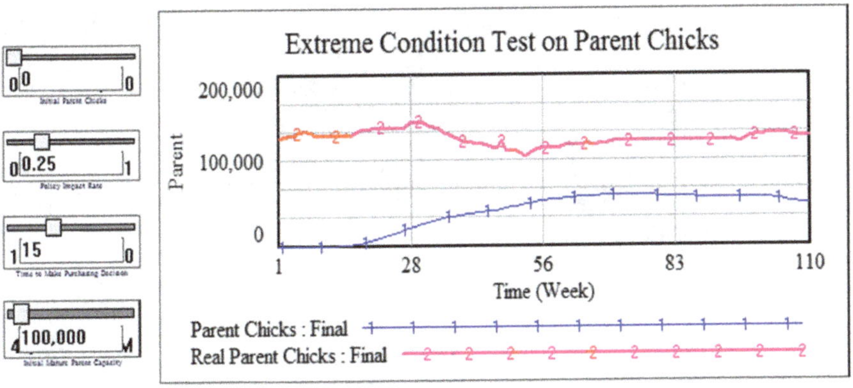

Fig. 4.5 Extreme condition test for parent chicks

4.9 Sustainable Poultry Production

This research creates a composite poultry supply chain model to ensure long-term viability. In reality, the majority of poultry processing methods are dispersed. The current study attempted to unite all fragmented processes under a single umbrella, with minor extensions, to develop an effective, long-term poultry production process. Figure 4.6 depicts a simplified framework for a long-term poultry supply chain, which was previously depicted in a more complex form in Fig. 4.2. Better input can aid in the production of high-quality output. The model was created using a real-world industry that is already using and implementing many scientific processes to achieve sustainability. Other supply chain processes have been incorporated into the case parent stock (PS) company's system to make the industry more efficient. They include a hatchery, a feed mill, a supply chain for day-old chicks, and a by-product processing unit. Notably, the length of the supply chain is determined by the scope of the poultry business operation, as there are numerous options. Grandparent, parent, ultimate farming, feed processor, feed supplier, chicks and chicken supplier, chicken processor, and all intermediary businesses are examples of operations. Poultry is a business that is completely reliant on input, and a given input will drive the supply chain operations that follow for a specific period. There is no other way to increase production or operation because the process is dependent on the number of chicks or eggs available. Reduced production can occur at any time as a result of natural disasters and calamities. As a result, the poultry operation's main output (chickens and eggs) is dependent on the given input, while the wastes generated are based on existing flock sizes.

All three factors (social, economic, and environmental) were directly benefited by poultry activities (Table 4.4). This triple bottom line success is defined as ensuring the long-term viability of a supply chain (Craig & Dale, 2008; Lee, 2004; Savitz & Weber, 2006). Processing by-products from poultry wastes, such as biogas, fertilizers, and fish feed, can benefit social, economic, and environmental. Recycling poultry

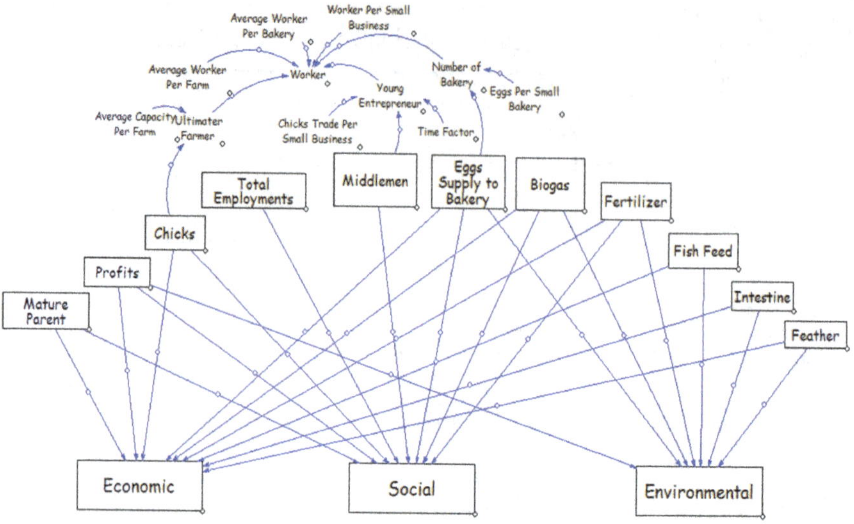

Fig. 4.6 Sustainable poultry process

wastes can help to protect the environment, create small businesses to process it, and generate additional revenue by selling by-products. Even if the waste processors use the by-products for their consumption rather than selling them, they save money, which is a direct economic benefit. The previous discussion of poultry activities and sustainability revealed that the case industry's activities covered sustainability theory and its components. Nonetheless, some of the operations managed by third-party companies were conducted unorganized and unscientific, which was a source of concern. If these operations or processes could operate as part of an integrated model, the industry would achieve all three aspects of sustainability.

Table 4.4 Benefits from poultry activities (Rahman, 2013a, 2013b)

No.	Main activity	Sustainable benefits
1	Rearing Parent Chicks, Mature Parent and Broiler birds	Social and Economic
2	Hatchery Operation and Producing Chicks	Social and Economic
3	Distributing Chicks, Middlemen Operation and Employ Agent and Sub-agent	Social and Economic
4	Farming Operation and Chicken House	Social and Economic
5	Processing By-Products from Poultry Wastes	Social, Economic, and Environmental

4.9.1 Economic Activities in Poultry Process

The main source of concern in achieving sustainability is economic gain. Therefore, the company is constantly working to achieve its maximum targeted profit to fulfill other responsibilities such as social and environmental responsibility. In the integrated poultry industry, three economic activities were observed: financial profitability (Dees, 1998), proper value addition (Cobb et al., 2009), and sales and cost of goods (Cobb et al., 2009; GRI, 2009). In practice, an integrated poultry supply chain increases profitability by lowering costs and increasing sales efficiency. Simultaneously, surplus poultry products are converted into value-added products that can be sold on the open market, ensuring the industry's long-term viability.

4.9.2 Social Activities in Poultry Process

One of the components of sustainability is social benefits, which are primarily a concern for society and the community. In GRI, AICHE, and Dow Jones reports, social indicators in sustainability concepts are listed. A sustainable poultry supply chain process can provide several direct and indirect social benefits (Shamsuddoha et al., 2021). The main identified social factors maintained through the poultry process are Employment Creation (Cobb et al., 2009), Poverty Reduction (Yunus, 2007), Create Self-employed Young Entrepreneurs (Åstebro & Thompson, 2011), and Creating New Ventures and Family Business Creation (Dyer & Chu, 2003). Parent Stock and broiler farming, hatcheries and middlemen, as well as other participants in the supply chain, keep these social activities going (distributor, agent, raw materials suppliers, and by-products processor).

The industry does not have to keep track of how it contributes to social issues. However, based on product turnover, people involvement, supply chain networks, and other factors, the government or non-government agencies can quickly assess this. For example, the poultry industry makes a significant contribution to both the rural and urban economies. This is because people participate in the supply chain network as workers, agents, and mediators, among other roles. Workers on participating farms, for example, can escape poverty by performing their jobs to a high standard. Furthermore, a large number of people are employed by supply chain members as workers. As a result, the case industry maintains 70% of the supply chain networks for distributing day-old chicks. Such a network expands the possibilities for eradicating poverty, particularly for many people living in rural areas.

4.9.3 Post-COVID Implications for Poultry Supply Chains

Covid-19 severely hits the poultry industry and its stakeholders. Bangladesh poultry industry and farmers have been struggling to generate a profit through their operations since Covid started. Before the pandemic, this sector was severely hit due to negative propaganda on poultry feed productions, quality assurance, and other macro influences like the economy, buying affordability, and competitive proteins availability at a lower price. For instance, hybrid Tilapia fish got massive attention from the poor and middle-class families due to cheaper protein sources. However, Covid-19 appeared at the beginning of 2019, and the poultry supply chain affects one more time with massive destructions. As a result, so many farmers lost their capital, stopped farming, and became unemployed.

Moreover, poultry products like day-old-chicks, meat, and eggs are not reaching the customers due to pandemic-related lockdowns and restrictions on roads and highways. Relevant transport services failed to reach farmers' doors, chicks, feed, medicine, and other utilities. Thus, the bad times continued for the farmers, middlemen, hatcheries, parent stock or breeder farmers, and grandparent farmers. Six to eight million people related to the poultry industry have fallen into uncertainties and they do not know when the situation will get normal for their business. Nevertheless, the case industry holds its nerve and continuous operation for the better tomorrow.

Poultry stakeholders are taken some steps to overcome the situation in the post-covid era which are in the following:

a. Keep growing poultry based on the demand they can meet practically.
b. Negotiate with the government to treat poultry fleet specially and not to consider as regular transportation. Such initiative succeeded, and the government allowed poultry transport under the covid lockdown.
c. Promote poultry meat and eggs which are the cheaper sources of protein and related protein intake with the health issues for the consumers. More protein will give them more immunity and it works.
d. Gradually increase the productions to push the market for more consumption.
e. Increase biosecurity measures for healthier chicken and eggs productions.
f. Incorporate modern technology like the robotic process in hatchery, farming, vaccination, and transportation.
g. Restore the supply chain for collecting raw materials from Europe, the USA, and South America.
h. Develop reverse loops for poultry wastes to save environments.
i. Creating more social impacts for society.

4.10 Conclusions

In summary, the concept of system dynamics simulation modeling is used in a wide range of industries worldwide. The ability of this methodology to provide futuristic behavior that aids decision-makers in preparing to act on time is a significant benefit. It now presents a significant opportunity for poultry entrepreneurs to maximize their profits by utilizing their potentiality. Integrated poultry management has the potential to benefit the economy, society, and environment. It can also help to create more job opportunities, expand the scope of the small and medium-scale industry, achieve social benefits, and keep our environment clean and sanitary. Sustainable poultry farming also comes with advantages of achieving various benefits. In addition to measuring quantity, the model output also includes associated consequences, which aid poultry stakeholders in making the best decision possible in a given situation. Future research could delve into the finer points of the entire industry operation and its efficiency.

References

Acs, Z., & Armington, C. (2004). Employment growth and entrepreneurial activity in cities. *Regional Studies, 38*, 911–927.

Ahmad, N., & Seymour, R. (2008). *Defining entrepreneurial activity: Definitions supporting frameworks for data collection.*

Ahuja, V., & Sen, A. (2007). *Scope and space for small scale poultry production in developing countries.* Indian Institute of Management.

Akter, S., & Farrington, J. (2007). *Sustainability of an innovation for poverty alleviation: The case of Bangladesh poultry model.*

Akter, S., & Farrington, J. (2008, July 8). Poverty transition through targeted programme: The case of Bangladesh poultry model. In *82nd Annual Conference of the Agricultural Economics Society.* Royal Agricultural College, UK. http://ageconsearch.umn.edu/bitstream/36765/2/Akter_farrington.pdf

Akter, S., & Farrington, J. (2009). What makes exit from poverty: Investigation of smallholder women livestock farmers in Bangladesh. In *2009 Conference, August 16–22, 2009.* International Association of Agricultural Economists, Beijing, China.

Alam, G. M., Hoque, K. E., Khalifa, M. T. B., Siraj, S. B., & Ghani, M. F. B. A. (2009). The role of agriculture education and training on agriculture economics and national development of Bangladesh. *African Journal of Agricultural Research, 4*, 1334–1350.

Asaduzzaman, M. (2000). Livestock sector, economic development and poverty alleviation in Bangladesh. In *Changing rural economy of Bangladesh* (pp. 42–53). Bangladesh Economic Association.

Åstebro, T., & Thompson, P. (2011). Entrepreneurs, Jacks of all trades or Hobos? *Research Policy, 40*, 637–649.

Balci, O. (1994). Validation, verification, and testing techniques throughout the life cycle of a simulation study. *Annals of Operations Research, 53*, 121–173.

Barlas, Y. (1996). Formal aspects of model validity and validation in system dynamics. *System Dynamics Review, 12*, 183–210.

Barratt, M. (2004). Understanding the meaning of collaboration in the supply chain. *Supply Chain Management: An International Journal, 9*, 30–42.

Barrera-Roldán, A., & Saldıvar-valdés, A. (2002). Proposal and application of a sustainable development index. *Ecological Indicators, 2*, 251–256.

BBS. (2010). *Rural poverty portal* [Online]. Dhaka. http://www.ruralpovertyportal.org/web/guest/country/statistics/tags/bangladesh. Accessed 29 Mar 2012.

Belisle, P. T. (2011). Sustainability/climate change. *Strategic Planning for Energy and the Environment, 30*(4), 71–78.
Bjerke, B. (2007). *Understanding entrepreneurship.* Elgar.
Butler, C. D. (2000). Global change hum. *Health, 1*, 156.
Chen, C.-N., Tzeng, L.-C., Ou, W.-M., & Chang, K.-T. (2007). The relationship among social capital, entrepreneurial orientation, organizational resources and entrepreneurial performance for new ventures. *Contemporary Management Research, 3*, 213–232.
Cobb, C., Schuster, D., Beloff, B., & Tanzil, D. (2009). The AIChE sustainability index: The factors in detail. *Chemical Engineering Progress, 105*, 60.
Cohen, L., Manion, L., & Morrison, K. (2011). *Research methods in education.* Routledge.
Cooper, C. M., Lambert, M. D., & Pagh, D. J. (1997). Supply chain management: More than a new name for logistics. *International Journal of Logistics Management, 8*, 1–9.
Coulthard, S., Johnson, D., & Mcgregor, J. A. (2011). Poverty, sustainability and human wellbeing: A social wellbeing approach to the global fisheries crisis. *Global Environmental Change, 21*, 453–463.
Council, S.-C. (1999). *Supply-chain council.* The Council.
Cox, J. F., Blackstone, J. H., & Spencer, M. S. (1995). *APICS dictionary* (8th ed.). American Production and Inventory Control Society.
Craig, C. R., & Dale, R. S. (2008). A framework of sustainable supply chain management: Moving toward new theory. *International Journal of Physical Distribution & Logistics Management, 38*(5), 360–387.
Cresswell, J. W. (2003). *Research design: Quantitative, qualitative & mixed method approaches.* Sage.
Daily, G. C., & Ehrlich, P. R. (1992). Population, sustainability, and Earth's carrying capacity. *BioScience, 42*, 761–771.
Davis, J. L., Bell, R. G., Payne, G. T., & Kreiser, P. M. (2010). Entrepreneurial orientation and firm performance: The moderating role of managerial power. *American Journal of Business, 25*, 41–54.
Davis, K. (1990). Population and resources: Fact and interpretation. *Population and Development Review, 16*, 1–21.
De la Fuente, M., Ros, L., & Ortiz, A. (2010). Enterprise modelling methodology for forward and reverse supply chain flows integration. *Computers in Industry, 61*, 702–710.
Dees, J. G. (1998). *The meaning of social entrepreneurship.* Comments and suggestions contributed from the Social Entrepreneurship Funders Working Group, 6pp.
Dolberg, F. (2004). *Review of household poultry production as a tool in poverty reduction with focus on Bangladesh and India.* National Dairy Development Board.
Dyer, J. H., & Chu, W. (2003). The role of trustworthiness in reducing transaction costs and improving performance: Empirical evidence from the United States, Japan, and Korea. *Organization Science, 14*, 57–68.
Elkington, J. (1994). Towards the sustainable corporation: Win-win-win business strategies for sustainable development. *California Management Review, 36*, 90–100.
Elkington, J. (2004). Enter the triple bottom line. *The Triple Bottom Line: Does It All Add Up, 11*, 1–16.
Ellram, L., & Cooper, M. (1993). Characteristics of supply chain management and the implications for purchasing and logistics strategy. *International Journal of Logistics Management, 4*, 1–10.
Fairclough, N. (2003). *Analysing discourse: Textual analysis for social research.* Routledge.
Forrester, J. W., & Senge, P. M. (1980). Tests for building confidence in system dynamics models. *TIMS Studies in Management Sciences, 14*, 209–228.
Freytag, A., & Thurik, R. (2010). *Entrepreneurship and its determinants in a cross-country setting.* Springer.
Goodland, R., & Daly, H. (1996). Environmental sustainability: Universal and non-negotiable. *Ecological Applications, 6*, 1002–1017.
GRI. (2009). *What is sustainability reporting.* Retrieved August 14, 2009.

Grönroos, C. (1997). Value-driven relational marketing: From products to resources and competencies. *Journal of Marketing Management, 13*, 407–419.

Hall, J. K., Daneke, G. A., & Lenox, M. J. (2010). Sustainable development and entrepreneurship: Past contributions and future directions. *Journal of Business Venturing, 25*, 439–448.

Hancock, G. (1992). *Lords of poverty: The power, prestige, and corruption of the international aid business*. Atlantic Monthly Press.

Heck, R. K., & Stafford, K. (2001). The vital institution of family business: Economic benefits hidden in plain sight. In *Destroying myths and creating value in family business* (pp. 9–17).

Homer, J., & Oliva, R. (2001). Maps and models in system dynamics: A response to Coyle. *System Dynamics Review, 17*, 347–355.

Honig, B. (1998). What determines success? Examining the human, financial, and social capital of Jamaican microentrepreneurs. *Journal of Business Venturing, 13*, 371–394.

Jain, V., Wadhwa, S., & Deshmukh, S. (2009). Select supplier-related issues in modelling a dynamic supply chain: Potential, challenges and direction for future research. *International Journal of Production Research, 47*, 3013–3039.

Jennings, P. D., & Zandbergen, P. A. (1995). Ecologically sustainable organizations: An institutional approach. *The Academy of Management Review, 20*, 1015–1052.

Johnson, R. B., & Onwuegbuzie, A. J. (2004). Mixed methods research: A research paradigm whose time has come. *Educational Researcher, 33*, 14–26.

Jørgensen, H. D. (2004). *Interactive process models*. Norwegian University of Science and Technology.

Knoepfel, I. (2001). Dow Jones sustainability group index: A global benchmark for corporate sustainability. *Corporate Environmental Strategy, 8*, 6–15.

Kocabasoglu, C., Prahinski, C., & Klassen, R. D. (2007). Linking forward and reverse supply chain investments: The role of business uncertainty. *Journal of Operations Management, 25*, 1141–1160.

Krantz, L. (2001). *The sustainable livelihood approach to poverty reduction: An introduction*. SIDA.

Kreiser, P. M., & Davis, J. (2010). Entrepreneurial orientation and firm performance: The unique impact of innovativeness, proactiveness, and risk-taking. *Journal of Small Business & Entrepreneurship, 23*, 39–51.

Larraâin, J. (1979). *The concept of ideology*. University of Georgia Press (Athens).

Lazear, E. (2003). *Entrepreneurship* (IZA Disc. Paper).

Lee, H. L. (2004). The Triple-A supply chain. In *The 21st-century supply chain*. Harvard Business Review.

Lummus, R. R., & Alber, K. L. (1997). *Supply chain management: Balancing the supply chain with customer demand*. APICS Educational and Research Foundation Inc.

Lummus, R. R., & Vokurka, R. J. (1999). Defining supply chain management: A historical perspective and practical guidelines. *Industrial Management & Data Systems, 99*, 11–17.

Lynch, R. G. (2004). *Economic, fiscal, and social benefits of investment in early childhood development*. Economic Policy Institute.

Mcmichael, A. J., Butler, C. D., & Folke, C. (2003). New visions for addressing sustainability. *Science, 302*, 1919–1920.

Munro, D. A., & Holdgate, M. W. (1991). *Caring for the earth: A strategy for sustainable living*. International Union for the Conservation of Nature and Natural Resources.

Norman, W., & Macdonald, C. (2004). Getting to the bottom of "triple bottom line". *Business Ethics Quarterly, 14*, 243–262.

Peacocka, C., & Shermanb, D. M. (2010). Sustainable goat production: Some global perspectives. *Small Ruminant Research, 89*, 70–80.

Qudrat-Ullah, H., & Seong, B. S. (2010). How to do structural validity of a system dynamics type simulation model: The case of an energy policy model. *Energy Policy, 38*, 2216–2224.

Quinn, F. J. (1997). What's the buzz? *Logistics Management, 36*, 43–47.

Rahman, M. R. (2013a, September 7). *RE: Modern poultry operation*. Type to Shamsuddoha, M.

Rahman, M. R. (2013b, February 1). *RE: Operation of Nahar Agro Complex Limited*. Type to Shamsuddoha, M.

Rahman, S. (2012, January 25). Poultry litter can ease energy crisis. *The Daily Star*.

Reynolds, P., Bygrave, W., Autio, E., & Cox, L. (2001). M. Hay, 2002. *Global entrepreneurship monitor, 2002 executive report*.

Rhyne, E. (1998). The yin and yang of microfinance: Reaching the poor and sustainability. *MicroBanking Bulletin, 2*, 6–8.

Saleque, M. A. (2013, June 12). *RE: Bangladesh poultry and its future*. Type to Shamsuddoha, M.

Sargent, R. G. (2005). Verification and validation of simulation models. In *Proceedings of the 37th conference on Winter simulation* (pp. 130–143). Winter Simulation Conference.

Savitz, A. W., & Weber, K. (2006). *The triple bottom line*. Jossey-Bass.

Seelos, C., & Mair, J. (2005). Social entrepreneurship: Creating new business models to serve the poor. *Business Horizons, 48*, 247–252.

Seelos, C., & Mair, J. (2007). Profitable business models and market creation in the context of deep poverty: A strategic view. *The Academy of Management Perspectives, 21*, 49–63.

Shamsuddoha, M., Quaddus, M. A., & Woodside, A. G. (2021). Environmental sustainability through designing reverse logistical loops: Case research of poultry supply chains using system dynamics. *Journal of Business & Industrial Marketing*.

Sikdar, S. K. (2003). Sustainable development and sustainability metrics. *AIChE Journal, 49*, 1928.

Sterman, J. D. (2000). *Business dynamics: Systems thinking and modeling for a complex world*. Irwin McGraw-Hill.

Sterman, J. D. (2001). System dynamics modeling. *California Management Review, 43*, 8–25.

Stock, J. H., & Watson, M. W. (2003). Has the business cycle changed and why? In *NBER Macroeconomics Annual 2002* (Vol. 17). MIT Press.

Sundin, E. (2011). Entrepreneurship and social and community care. *Journal of Enterprising Communities: People and Places in the Global Economy, 5*, 212–222.

Svensson, G. (2007). Aspects of sustainable supply chain management (SSCM): Conceptual framework and empirical example. *Supply Chain Management, 12*, 262–266.

Tilman, D., Cassman, K. G., Matson, P. A., Naylor, R., & Polasky, S. (2002). Agricultural sustainability and intensive production practices. *Nature, 418*, 671–677.

Tilman, D., Wedin, D., & Knops, J. (1996). *Productivity and sustainability influenced by biodiversity in grassland ecosystems*.

Venkataraman, S. (2002). *The distinctive domain of entrepreneurship research*. Edward Elgar Press.

Wagner, J. (2003). Testing Lazear's jack-of-all-trades view of entrepreneurship with German micro data. *Applied Economics Letters, 10*, 687–689.

Wakeford, J. J. (2012). *Socioeconomic implications of global oil depletion for South Africa: Vulnerabilities, impacts and transition to sustainability*. Stellenbosch University.

Ward, L. (2002). *Environmental policies for a sustainable poultry industry in Sussex County*. University of Delaware.

WCED. (1987). *Our common future* (The Brundtland Report). Oxford University Press.

Wolstenholme, E. F. (1990). *System enquiry: A system dynamics approach*. John Willey & Sons.

Yunus, M. (2007). *Creating a world without poverty: Social business and the future of capitalism*. Public Affairs Store.

Zebda, A. (2002, September). Using cost-benefit analysis for evaluating decision models in operational research. *Journal of American Academy of Business, 2*, 106–114.

Chapter 5
Environmental Voluntary Programs in the Transport Industry in the Post-COVID Situation: The French Experience

Enzo Bivona and Gisele Mendy Bilek

Abstract This study investigates the drivers impacting on the success of the environmental voluntary program recently launched in France. This program, titled EVE (Engagements Volontaires pour l'Environnement), aims to improve transport operators' efficiency and to reduce CO_2 emissions in the environment. Due to the complexity of the context in which the EVE program is implemented, we used a System Dynamics (SD) approach. This study offers multiple contributions. First, studying the dynamic interdependences of voluntary programs aimed at reducing carbon emissions in the transport industry can help policymakers in designing successful policies. In addition, the study can help transport operators engaged in environmental voluntary programs to build a durable competitive advantage, while complying with environmental policies, particularly in the post-COVID situation.

Keywords System dynamics · Global North · GHG emissions · Stakeholders · Transport services · Supply chain disruptions · Economic development · Medical supplies · Economic consequences · Feedback loops · Stock-and-flow structure · Group-model building · Global economy · Carbon emissions · Energy efficiency · Voluntary programs · France · Environmental impacts · Public and private actors · Environmental initiatives · Operators' engagement · Institutional theory · Sustainable behaviour · Technological evolution · Coercive isomorphism · Trading alliances

---

E. Bivona (✉)
Business Management, University of Palermo, Palermo, Italy
e-mail: enzo.bivona@unipa.it

G. M. Bilek
Business Management, IAE Pau Bayonne, LIREM, Université de Pau Et Des Pays de L'Adour, Pau, France
e-mail: gisele.mendy-bilek@univ-pau.fr

H. Qudrat-Ullah (ed.), *Understanding the Dynamics of New Normal for Supply Chains*, Understanding Complex Systems,
https://doi.org/10.1007/978-3-031-07333-5_5

5.1 Introduction

While countries in the Global North are responsible for building up most of the Greenhouse Gas (GHG) emissions in the atmosphere, some have managed to reach a peak in emissions and sustain long-term reductions (Le Quéré et al., 2019; Minx et al., 2021; William et al., 2022). To pursue the Paris Agreement, which implies a steep reduction towards the net-zero global emissions within the next few decades (IPCC, 2018), France have launched, among the others, a voluntary environmental program in the transport sector. It is worth remarking that transport is the largest emitting sector in France, and it accounts for 40% of country's GHG emissions, 30% of which are due to freight mobility (Climate Transparency, 2021). With this intent to pursue the French National Low Carbon Strategy, which aims to pursue carbon neutrality by 2050, similarly to other countries, in which successfully environmental voluntary programs are currently in place (see, for instance, SmartWay in the USA), the national government launched the program EVE (Engagements Volontaires pour l'Environnement).

The EVE program's goal is twofold. On the one side, it aims to improve transport operator's efficiency. On the other side, it pursues a reduction in GHG emissions in the environment. The EVE program matches together multiple private and public stakeholders. However, the multiple relationships between public and private actors involved in the program (ranging from the program coordinator to transport organizations), the complexity of the supply chain in the transport and logistics sector (including the differences among the transport operators sub-groups), and the deep crisis generated by the COVID-19 pandemic may prevent the achievement of the desired project outcomes.

The recent COVID-19 coronacrisis has produced restrictions on freedom at both individual movement and commercial exchange of goods (Kazancoglu et al., 2022; Kumar & Singh, 2021). In this unprecedented situation, the transport sector demonstrated that it is essential to social and economic development and guarantees mobility across countries. It ensures the supply of essential goods (from food to medical supplies), as well as a large range of raw materials upon which both businesses and consumers depend on. During the spread of COVID-19, many governments have introduced restrictions on domestic and border crossing transport services. This generated a downturn in trade and supply chain disruptions impacting on the economic development of a large number of countries. Although transportation firms contributed to keep supermarket shelves stocked with essential goods and deliver medical supplies, COVID-19 restriction policies (e.g., border closures, export restrictions, social distancing, lockdowns and closures of non-essential sectors) generated, particular, in those of small-medium dimension, a rapid deterioration of the liquidity and profitability exposing them at high risk of bankruptcy.

While the economic consequences caused by the COVID-19 pandemic have not yet been fully assessed, this new scenario highlighted the central role of mobility on GHG emissions (Aktar et al., 2021). To achieve carbon neutrality by 2050, the call for more interventions on mobility will no doubt remain acute despite technological advances, such as the diffusion of electric vehicles.

To investigate the system complexity in which the EVE program is implemented, we suggested to use the System Dynamics (SD) methodology (Forrester, 1961; Sterman, 2000). The use of a SD model can support the EVE program coordinator to design and assess effective policies to pursue the expected goals. As the research project is in the early stage, this study outlines the research path, main feedback loops, and a preliminary stock-and-flow structure to be used in group-model building sessions with project' participants.

This study offers multiple contributions from theoretical and practical points of view. First, studying the dynamic interdependences of voluntary programs aimed at reducing carbon emissions in the transport industry can help policymakers in designing successful policies. This research has also important implications for managers. The study can help transport operators engaged in environmental voluntary programs to build a durable competitive advantage, while complying with environmental policies, particularly in the post-COVID situation.

5.2 Relevance of the Transportation and Logistics Sector in CO_2 Emissions: Environmental Mandatory and Voluntary Programs

Trade growth and expanding global economy are creating an endless demand for freight transport capacity and infrastructure. As a result, carbon emissions from freight transport are growing at a rapid rate. Thus, projections of carbon emissions by 2050 from global freight could nearly quadruple. In Europe and throughout the world, road transportation is the largest source of greenhouse gases from freight (OECD/ITF, 2015). Looking beyond 2020, in its climate and energy policy framework for 2030, the European Union set itself a target of reducing emissions to 40% below 1990 levels by 2030. Transport and logistics activities alone account for more than 20% of global CO_2 emissions (Davydenko et al., 2014).

Several public and private initiatives are thus deployed to encourage the actors to collaborate and to instil a policy of "decarbonisation" in particular in the sector of transport and logistics. The need to implement a mode of sustainable development, combining economic, social and environmental development is today widely recognised. Improving the energy efficiency of road transport is more than ever essential to achieve the objectives of a low carbon strategy at the country level.

Since the late 1990s, to face the carbon emissions reduction in the transport sector two different, although complementary, approaches emerged in the forms of mandatory program, regulated by specific legislations, and voluntary programs, engaging multiple actors from the public and the private sphere.

Mandatory programs in the transport sector are implemented in a few countries like UK, through the introduction of the reporting on Carbon Footprint (Dadhich et al., 2015), and France, with the CO_2 reporting. These regulations and legislations either provide an incentive or impose a great pressure on companies to adopt green and sustainable practices and collaborations along the supply chain. In particular, in the transport for freight and passengers in France, a regulatory device was set up since October 2013 bearing on the obligation of calculating and reporting carbon emissions for every transport services having a point of origin or destination on the French territory (decree 2011–1336). Carbon reporting in transport services aims to improve information given to buyers to encourage them to reconsider their choice regarding the design of their supply chains. A research conducted on these mandatory programs in France (Mendy Bilek et al., 2017) confirms the importance of such a regulatory systems. However, research findings also remarked that they do not seem sufficient to drive, alone, a real change in the system.

Since the beginning of the year 2000, several voluntary programs for measurement and reduction of carbon emissions are implemented in US and in Europe. In 2004 in US, the EPA launched Smartway Transport Partnership a public–private initiative between freight shippers, carriers, logistics companies and others stakeholders to voluntarily improved fuel efficiency and reduce environmental impacts from freight transport (Bynum et al., 2018; Tan & Blanco, 2009). In Europe similar programs are in The Netherlands and in France. The Dutch "Lean and Green Program" encourages partnership between shippers, carriers and technology providers to measure and improve CO_2 emissions. In France a similar voluntary program is in place since the year 2009 for carriers, through the framework "Objectif CO_2 les transporteurs s'engagent". More recently, in 2016 the program FRET 21 includes shippers. Wolmarans et al. (2014) show that shipper initiatives are largely driven by company policy and that shippers tend to push sustainability requirements onto the carriers that work for them. Also, carriers are motivated to adopt sustainable business practices that will make them more competitive and help reduce costs.

However, the lack of uniform assessment and reporting mechanisms greatly reduced its value for either shippers or carriers to influence decisions (Bynum et al., 2018). Furthermore, the multiple relationships between public and private actors involved in the program (ranging from the program coordinator to the transport organisations), the complexity of transport and logistics sector (including the differences among the transport operators sub-groups and the resistance of transport firms to introduce innovative practices impacting on the environment) may prevent the achievement of the desired project outcomes. The EVE (Engagements Volontaires pour l'Environnement) program recently launched in France does not constitute an exception. This program aspires to improve transport operators' efficiency and to reduce the impact of transportation flows on the environment. It is coordinated by a public agency and it targets more than 700 carriers, 200 shippers and 70 freight forwarders.

Due to the level of complexity characterising such an environmental voluntary program, the "French Environment & Energy Management Agency" ADEME (Agence de l'Environnement et de la Maîtrise de l'Énergie), who supervises the EVE project, financed a research project aimed at designing a collaborative model to effectively implement the environmental voluntary program EVE. This research project strongly relies on the findings that would emerge from the System Dynamics modelling of the EVE program. The analysis of the context in which all stakeholders involved operate and the investigation of the transport operators' sub-groups profiles will offer the ground to model the processes underlying actor's program engagement and the implementation of transport operator efficiency and environmental initiatives. Furthermore, by designing appropriate key performance indicators (such as program attractiveness, transport operators' engagement, and program effectiveness), the SD model will support the EVE coordinator to test alternative policies and to assess the success of the program. As the research project is an early stage, this chapter outlines the research path and the preliminary stock-and-flow structure that will be adapted based on the project' participants' viewpoints during the group-model building sessions.

The structure of the chapter is articulated as follows. The next section analyses transport organizations' behaviors in environmental programs. Section 5.4 describes the context of sustainable transportation policies in France and introduces the EVE environmental voluntary program. Section 5.5 presents and discusses the research approach used to build the SD model. This section also portrays the main feedback loops and the preliminary stock-and-flow structure built in the early stage of this research. Section 5.6 offers some concluding remarks.

5.3 Framing Transport Organization's Behaviors in Environmental Programs: The Neo-Institutional Theory Perspective

The neo-Institutional theory provides a useful theoretical framework for research in sustainable transportation to explain how the external factors push organizations to implement environmental practices in their supply chain management (Sarkis et al., 2010). Several studies have examined its important influence on the firms' performance (Tate et al., 2012; Zhu & Sarkis, 2007), and it has been proved to have a positive influence on firms' sustainable practices implementation (Chu et al., 2017). The key components of the institutional theory are the three mechanisms of isomorphism, identified by Di Maggio and Powell (1983), which are *coercive, normative,* and *mimetic isomorphism.* They mention that "isomorphism is a constraining process that forces one unit in a population to resemble other units that face the same set of environmental conditions" (Dimaggio & Powell, 1983). In the context of sustainable behavior, isomorphism is interpreted to the external pressures that lead organizations to adopt similar structures or strategies in supply chain management to respond to social expectations and achieve sustainable development (DiMaggio & Powell, 1983; Chu et al., 2017). For example, external pressures may include changes in cultural or social values, technological evolution, regulations (Glover et al., 2014; Sayed et al., 2017). Accordingly, these pressures can be grouped into three categories *coercive pressures, normative pressures, and mimetic pressures.*

According to Di Maggio and Powell (1983), coercive isomorphism derives from the political influences of the problem of legitimacy. Organizations confront the coercive pressure resulting from other organizations they rely on (e.g. government agencies, headquarters) and the cultural expectations in society (e.g. legislation, social norms, and standards). The rationale of the voluntary information disclosure program is to provide better information to stakeholders, customers, employees, government agencies, and NGOs to constitute a form of institutional pressure that can motivate firms to improve along with metrics and measures the information disclosed.

Many empirical studies focused on how organizational practices diffuse through an organizational field but few investigations try to understand the conditions under which institutional pressure and organizational characteristics explain the adoption of compliance strategies (Delmas & Toffel, 2008).

Normative pressure stems from the professionalization and expectations relating to how work should be done professionally (DiMaggio & Powell, 1983). It may come out from the broader communities (e.g. markets, media, and the general public) (Zhu, 2016). Especially, external stakeholders who have direct or indirect interests in the organizations' environmental management, exert normative pressures that are regarded as legitimate for organizations within their industrial community (Sayed et al., 2017). Also, organizations are confronted with normative pressures that are exerted by sustainable trading alliances and associations that have the desire to work with them (Tate et al., 2012). Therefore, normative pressure can be perceived as an important driver for organizations to be more environmentally aware and respond to environmental issues to comply with social obligations (Glover et al., 2014).

Concerning *mimetic pressure*, there are, in most cases, proactive leaders, reactive followers, and stagnant laggards in any industry and any field of business. According to DiMaggio and Powell (1983), "organizations tend to model themselves after similar organizations in their field that they perceive to be more legitimate or successful". They seek to replicate the successful path of the leaders with the purpose to gain more benefits from the market (Prajogo & Olhager, 2012). In the context of institutional pressure, when industry leaders take an action in response to institutional pressure, the followers may simply follow suit as they perceive them to be more legitimate and successful, irrespective of whether they are directly affected by the regulation or not (Loannou & Serafeim, 2017). Firms that are uncertain of the external environment or incapable of interpreting institutional pressures on their own are quite likely to be influenced by mimetic isomorphism (DiMaggio & Powell, 1983). There is a prerequisite for mimetic isomorphism to take effect, which is that there must exist successful firms that can be imitated. This phenomenon (DiMaggio & Powell, 1983) explains the tendency to homogenize organizational behavior by identifying three mechanisms, including coercive isomorphism resulting from formal and informal pressure exerted for example by the state. But, as Meyer and Rowan (1977) point out, there may be a contradiction between compliance with the institutionalized rules and the search for efficiency required by internal coordination and control of activities. The decoupling between the obligation to display CO_2 and the requirements of profitability of the company sometimes undermined by the cost (real or perceived) of sustainable practices can thus generate contradictory tensions (Abernethy & Chua, 1996).

The results thus suggest that the evolution of practices (CO_2 display and/or CO_2 reduction) results in more from a combination of coercive, mimetic, and normative pressures. Coercion by customers and other stakeholders such as shareholders appears here to be potentially more effective than that exercised by the state. Companies would also tend to imitate (or "benchmark") those that have already proven themselves in practice. Finally, the development of benchmarks, labels, standards related to the measurement of CO_2 emissions within the profession could encourage companies to achieve greater compliance.

5.4 Designing a Sustainable Transportation Policy in France: The Expected Contribution of the EVE Environmental Voluntary Program

5.4.1 Designing a Sustainable Transportation Policy in France

In recent years, the French government and public authorities have been devoting all their attention to the reduction of CO_2 emissions, considered essential to deal with global warming. Several public and private initiatives are thus deployed to encourage the actors to collaborate and to instill a policy of "decarbonization" in particular in the sector of transport and logistics. The need to implement a model of sustainable development, combining economic, social, and environmental development, is today a consensus. In this perspective, improving the energy efficiency of road transport is more than ever essential to achieve the objectives set in France's low carbon national strategy. These objectives seem to be achievable only if all players in road transport are committed to improving their energy performance.

From the beginning of the 2000s, in France, the agency of Environment and Energy (ADEME) showed the need to adapt the transport and logistics sector to meet this challenge through the reorganization of production and purchasing systems aimed at improving both environmental performance and economic competitiveness. In particular, it encouraged some players in the supply chain, notably shippers (manufacturers and distributors) and carriers, to work together to control their environmental and energy performance. In 2008 started the "Objective Charter CO_2" program in road freight transport and the implementation of the "CO_2 Objective Label" (a certificate showing the adherence of the transport organization to comply with the program measures). In 2001, this program was extended to passengers' transport and in 2017 also to shippers (FRET 21 program).

The quantitative objectives set were well achieved since 5,500 companies were made aware, out of a target of 3,000, and almost 1,600 were supported. In 2016 and 2017, 540 charters were signed and more than 300 companies certified.

Can these results make it possible to establish a real virtuous dynamic in a sector that counts more than 6,000 passenger road transport companies and 35,000 road freight transport companies?

Unfortunately, the answer isn't positive. This appears evident if we consider that 80% of companies are made up of very small businesses, while most companies currently "labelled" have more than 50 employees.

Several reasons can explain the inertia of the sector despite the significant resources that have been deployed. At the structural level, the lack of incentive from prime contractors and the difficulty for companies to promote the process internally and at the commercial level are regularly highlighted. Also, the multiplication of systems and public and private initiatives without real coordination has given professionals the demotivating impression of a certain inconsistency. At a cyclical level, relatively low energy prices over the 2016–2017 period, and the economic difficulties of the sector have also limited the willingness of companies to commit to a long-term approach that requires immediate human and financial investment against future fallout.

Previous research (Mendy Bilek et al., 2017) confirmed that "the regulatory and voluntary mechanisms do not seem sufficient to stimulate, on their own, a real dynamic of change". For the above reasons, it is important to study the role and the interdependencies between the multiple stakeholders (e.g., public agencies, private transport organizations) involved in the environmental voluntary program and how to stimulate a virtuous interactive behaviour.

The above complex and dynamic picture outline a suitable field of study on which to apply the System Dynamics methodology (Forrester, 1961; Sterman, 2000). The SD methodology aims at supporting decision makers learning processes to better understand how to deal with complex phenomena (Sterman, 2000). Delays, nonlinearities, and policy resistance factors often make, public and private, managers' decision-making processes uncertain and investigated phenomena behaviors hard to interpret. Using feedbacks structures and simulation models, SD has shown its ability to support decision makers in dealing effectively with this level of complexity (Kunc & Morecroft, 2010; Rahmandad, 2015; Repenning, 2000; Sterman, 1989, 2000). As Sterman (2000) argues, particularly, simulations can be a very effective way to learn in and about complex systems. Feedback structures, i.e. closed cause-and-effect relationships between two or more variables, are considered responsible for the dynamic behavior portrayed by a given problem. In other words, the SD methodology tends to look inside a system for the real causes of the investigated phenomenon.

5.4.2 The French EVE Environmental Voluntary Program

The EVE program, from the French "Engagements Volontaires pour l'Environnement" (Voluntary Commitment to the Environment), attempts to improve transport operator's efficiency and to reduce Greenhouse Gas (GHG) emissions in the environment. This program can be associated with those initiatives oriented to pursue the French National Low Carbon Strategy, which aims for carbon neutrality by 2050. With this intent, the EVE program is promoted by the French ministry of the ecological and solidarity transition (e.g., the Ministry in charge of transportation), and it is funded using energy savings certificates financed by the Total Marketing France.

This program, similarly, to other countries, in which successfully environmental voluntary programs are currently in place (see, for instance, SmartWay in the USA), matches together different groups of private and public stakeholders.

In this case, the "French Environment & Energy Management Agency" ADEME (Agence de l'Environnement et de la Maîtrise de l'Énergie), supervises the EVE project, which is coordinated and implemented by a non-profit organisation, the "Eco CO_2". The Eco CO_2 covers different activities. First, it advertises and promotes the program goals with the intent to engage a high number of freight industry operators. The EVE program targets more than 700 carriers, 200 shippers, and 70 freight forwarders.

Second, it coordinates primary French freight professional organizations (such as AUTF, CGI, FNTR, FNTV, OTRE, and Union TLF), who are also partners of the EVE program, in designing effective energy savings and emission reductions measures. Typical examples are the development of fuel-saving technologies and the use of tracking tools to monitor efficiency improvement and emissions reduction.

Finally, Eco CO_2 also provides the "Objectif CO_2" certification to transport operators who comply with the suggested measures.

The expected success of the EVE program can lead to a win–win situation for the freight industry and the ADEME. On the one side, the freight industry can benefit from the support of experts in achieving fuel savings thereby making the sector more competitive. The "Objectif CO_2" certification can also help transport operators to improve their image and to meet customers' expectations, who are particularly sensitive to select suppliers adopting GHG emissions practices.

On the other side, the EVE program can support the ADEME to pursue the French National Low Carbon Strategy, as a result of the drop in fuel consumption and CO_2 emissions.

However, the multiple relationships between public and private actors involved in the program (ranging from the program coordinator to the transport professional organizations), the complexity of the transport and logistics sector (including the differences among the transport operators sub-groups and the resistance of transport firms to introduce innovative practices impacting on the environment) may prevent the achievement of the desired project outcomes. To support the EVE coordinator in designing and assessing effective policies to pursue the expected program goals, the use of an SD model is here suggested.

5.5 A Preliminary SD Model to Support the Design and Implementation of a Sustainable Transportation Policy in France

5.5.1 The Research Approach Used to Build the System Dynamics Model

The research lasts a total of 24 months, while the modeling phase covers about 14 months. In the initial stage (6 months), project activities are oriented to conduct the literature review of environmental voluntary programs and those factors facilitating or tackling the introduction of fuel-saving and GHG emissions measures in the freight industry. After this literature review, field research will be conducted with freight industry operators to investigate the level of participation in the EVE program, the obstacles, and the benefits recorded. The above findings will offer the basis to build a preliminary system Dynamics (SD) model (Forrester, 1961; Sterman, 2000). The SD methodology aims at supporting decision makers learning processes to understand better how to deal with complex phenomena. Very often, delays, nonlinearities and policy resistance factors make managers' decision-making processes uncertain and investigated phenomena behaviours hard to interpret. Through the use of feedbacks structures and simulation models, SD has shown its ability to support managers in dealing effectively with this level of complexity in multiple context (Bivona & Ceresia, 2008, Bivona & Montemaggiore, 2010; Bivona et al., 2019; Harrison et al., 2020; Kunc & Morecroft, 2010; Rahmandad, 2015; Repenning, 2000; Sterman, 1989, 2000). Feedback structures, i.e. closed cause-and-effect relationships between two or more variables, are considered responsible for the dynamic behaviour portrayed by a given problem. In other words, the SD methodology tends to look inside a system for the real causes of the investigated phenomenon.

However, such a richness of studies using dynamic approaches cannot be observed investigating environmental voluntary programs (for an exception, see Tan & Blanco, 2009). Therefore, this research aims at providing an additional study in this last direction.

Once the preliminary SD model will be built, then it will be used to conduct group-model building sessions with ECO CO_2 and freight industry operators' managers.

The SD literature remarks on the interactive nature of the model-building process (Richardson & Pugh, 1981; Roberts et al., 1983; Sterman, 2000; Vennix, 1996). This interaction can be detected at two different levels: among the multiple model building stages and, between the different actors involved, such as the modeler/s and final user/s (i.e., who will benefit from the model use).

The first level refers to the modeling process, which can be summarised in five recurrent steps (Sterman, 2000). The first step is the *problem articulation*, which dictates the boundary and the scope of the modeling effort. Once the problem is identified and observed over an extended time horizon to capture its potential symptoms, a *dynamic hypothesis* is formulated. This second step explains the problem in terms of the underlying feedback and stock-and-flow structure. The third step is the *model formulation*, which implies data collection and the estimation of parameters were not available or easy to access. The fourth step is *model testing*, consisting of the evaluation of the correct formulation and the robustness of the model to simulate the actual behavior of the investigated phenomenon. Finally, once the model is tested, it can be used for *policy design and evaluation* to intervene on the problematic behavior under investigation. Insights generated from simulation results can lead to a redesign of the feedbacks and the stock-and-flow structure, thereby changing the quality of information available and the adopted policies. Such an interaction is likely to feed both modeller and decision maker's learning processes.

The second level of interaction is particularly critical as the user/client cooperates with the modeler in providing information and data needed to feed all the modelling steps. As quantitative and, particularly, qualitative data (e.g., user's perceptions of the relevant feedbacks causing the observed phenomenon) characterise all stages of the modelling process, a more formal approach to collect, store and analyse data is required (Luna-Reyes & Andersen, 2003; Vennix, 1996). In particular, Vennix (1996) argues that the carefully design of the group model building process can contribute to successfully increase the effectiveness of the model, thereby enhancing team learning, fostering consensus and creating commitment with the outcomes.

The group model building process appears particularly suitable with the complexity outlined by the EVE project goals (reduction of fuel consumption and CO_2 emissions) and the presence of multiple actors involved in the process (ADEME, ECO CO_2, freight operators and freight professional organisations). Vennix (1996) suggests two alternatives to start the modelling process once the scope of the project is defined. The modeller can build the SD model from scratch involving main participants to offer their viewpoints or alternatively can construct a preliminary model which serves as starting point for the group-model building sessions.

In the first case, the SD model is built directly with participants in a group setting. However, if the written material, such as the literature review and project reports, is not adequate to build the preliminary model, the modeller can conduct in advance a number of interviews to get a better understanding of the problem.

In the second case, where project documentation is available and interviews with participant possible, the modeller can build a preliminary SD model. The model is then used to facilitate a discussion with key-actors involved in the project. Suggestions and comments from participants are used to fit the model with the participants' viewpoints.

In this research, we decided to build first a preliminary SD model based on project documentation. Then, the model will be validated during the group model-building sessions with project participants. This decision, as also suggested by Vennix (1996), would allow us to speed up the group model-build process and to allocate more time to data gathering, to fine tuning the model and to design and implement alternative policies. As the research project is the early stage, the remaining part of the chapter outlines the main feedback loops and the preliminary stock-and-flow structure that will be adapted based on project' participants viewpoints during the group-model building sessions.

5.5.2 Main Feedback Loops

Since the EVE initiative is a voluntary program, it needs to be attractive for shippers and carriers. Their participation in the program and their effective engagement are a prerequisite to collaborate with ECO CO_2 selected experts in identifying appropriate solutions to address energy and environmental issues in the transportation and logistics industry.

From the analysis of the literature on voluntary program it emerges the key role of stakeholders' awareness towards the adoption of environmental measures by transport organisations (shippers and carriers) in their operations. In fact, if consumers (e.g., shippers, in this case) are highly sensitive to environmental performance in the selection of the freight operators (e.g., carriers), this will make a pressure on carriers to enrol in initiatives, such as the EVE program. The enrolment of carriers in the EVE program can be also stimulated by advertising initiatives aimed at promoting participants benefits. This phenomenon can be associated to the reinforcing loop "R1 - *Carriers growth*" reported in Fig. 5.1. It is worth noting that the strength of such a reinforcing loop could be reduced by the effect of the environmental dynamism (Forliano et al., 2022; Romme et al., 2010) generated by the high uncertain produced by the diffusion of the COVID-19 pandemic. This unprecedented situation led multiple governments to introduce restriction policies, such as, export restrictions, social distancing and lockdowns, which generated a downturn in trade and supply chain disruptions impacting on the development of the economy (Kazancoglu et al., 2022; Kumar & Singh, 2021). Therefore, this may expose, particularly, small and medium size transport firms to a high risk of bankruptcy and prevent them to adhere in environmental voluntary programs, such as the EVE program, due to extra expenditure to improve their fleet efficiency, thereby reducing CO_2 emission.

Despite the possible constraints generated by the diffusion of the COVID-19 pandemic, an increase in the number of carriers enrolled in the EVE program can contribute to boost program attractiveness. In fact, the enlargement of the carriers enrolled leads to a higher number of new carriers who join the program. This is likely to expand also the professional experts engaged in the program. The role of such experts consists in supporting carriers to design and implement fuel-saving and emission reduction measures. Professional experts engaged in the program may show a similar behaviour to carriers. The raise in program attractiveness and incentives offered to professional experts can bring inside the program more experts, thereby expanding the number of professional experts engaged in the program (see feedback loop R2—*Professional experts growth*).

The increasing in carriers and professional experts may lead to two other reinforcing feedback loops. The loop "R3—*Effect of the program benefits on carriers dynamics*" is particular important. In fact, thought the participation in the program is an important indicator of its attractiveness, carriers implemented measures aimed at improving fuel saving and emission reduction captures the program effectiveness. It refers to the ability of the program to timely meet carriers' requests. This can lead to a boost in program effectiveness and perceived carriers' competitiveness, thereby

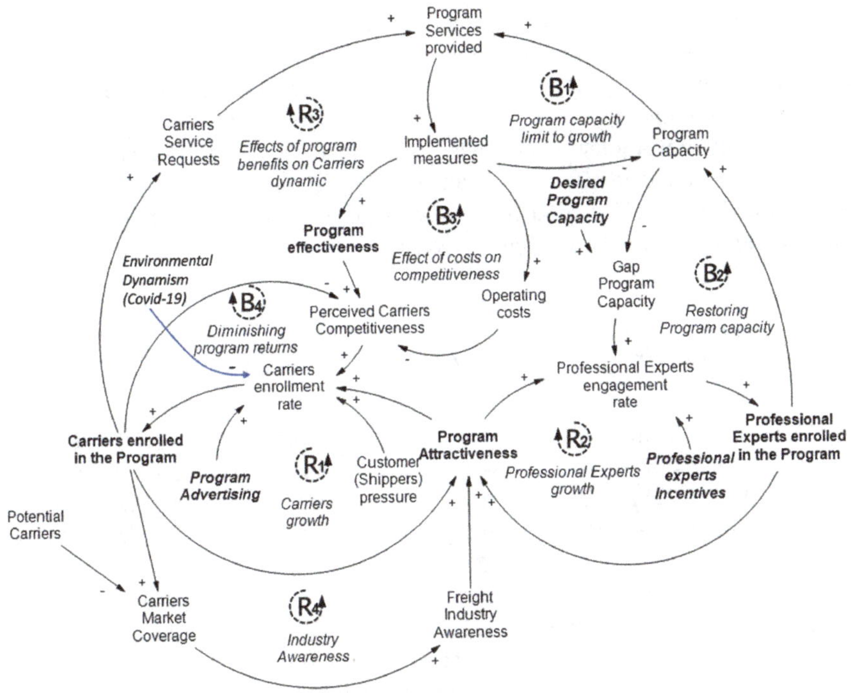

Fig. 5.1 Main feedback loops of the EVE program

bringing more carriers into the program. The causal loop "R4—*Effect of industry awareness*" results from the diffusion of the freight industry awareness. The increase in carriers can make the freight industry more aware of the potential benefits of the program and stimulate emulating behaviour by other carriers.

Figure 5.1 also depicts four balancing feedback loops which may contribute to limit or stabilise the EVE program desired results. As the number of carriers grows up, more services and implemented measures will result. Therefore, the EVE coordinator may experience a lack of available capacity limiting the expansion of program (see the balancing feedback loop B1—*Program capacity limit to growth*). This phenomenon can be counterbalanced by introducing a desired level of the program capacity. In such a way, the EVE coordinator through the use of professional experts incentives can stimulate the minimise the gap in program capacity, thereby restoring the desired level of services offered to carriers (see the balancing feedback loop B2—*Restoring desired program capacity*). Two other balancing loops may prevent the program to achieve the expected results. The balancing feedback loop "B3—*Operating costs increase*" shows how the investments sustained by carriers to implement the suggested measures may discourage transport operator to enrol in the program as perceived competitiveness slows down. This is particular true in the very highly competitive and uncertain transport industry. Many carriers are often sceptical to invest in innovation and technology which may not improve performance or may result in breakdowns and loss of service in the short term. Rather, they prefer to avoid innovation and pass the cost of inefficiency on to the final customers (e.g., shippers) via fuel surcharge policy (Bynum et al., 2018; Wolmarans et al., 2014). Though this may appear as a myopic policy, it may prevent carriers to incur in an economic loss in the short period. Another phenomenon that may tackle program benefits refers to the decreasing appealing it may encounter as soon as it is not perceived by carriers as a distinctive source of competitive advantage (see the balancing feedback loop B4—*Perceived Diminishing program returns*).

5.5.3 The Preliminary Stock and Flow Structure

As discussed in the previous section, based on the adopted research design, a preliminary stock and flow structure was built. Such a model will be used during the group-model building sessions with project' participants and it will be adapted to reflect their viewpoints. The preliminary model aims to capture the main reinforcing and balancing loops described in Fig. 5.1.

To build the stock and flow structure the concept of a multi-sided digital platform (Eisenmann et al., 2011; Ruutu et al., 2017) is here used. Similarly to the EVE program, multi-sided digital platforms aim at connecting demand-side (e.g., carriers) and supply-side (e.g., professional experts) participants through innovative forms of value creation processes (Täuscher & Laudien, 2018). When a community of actors is developing platform-based services, such as the EVE program, it is important that a critical mass of actors is reached in order to achieve self-sustaining growth.

Initially, platform development may be promoted or subsidized using external funding, but over the long term the success of a platform depends on its ability to attract customers. In the initial phases, the so-called 'chicken-and-egg' situation has to be faced. Too few demand- and supply-users inhibit the growth of each side of the user customer base, and vice versa (Casey & Töyli, 2012). If fact, if the number of carriers enrolled in the program is too low, professional experts will not join the program as well, resulting in a failure of the program.

The two sides of the program is modelled by extending the Bass (1969) model of innovation diffusion that considers adoption through exogenous efforts (advertising or incentives) and adoption from word-of-mouth (see Fig. 5.2). Here, the stocks of potential carriers and carriers are calculated separately from the stocks of potential professional experts and professional experts. The Carrier adoption rate depends on the advertising, customer (shippers) pressure and carriers program attractiveness, while professional experts engagement rate refers only to incentives. The model also includes discard rates, which depend on the carriers perceived program effectiveness and professional experts program attractiveness respectively. The program service capacity plays as important role in affecting carriers perceived program capacity. In fact, if the program shows a lack professional experts to adequately support carriers, carriers perceived program capacity declines leading to raise in the carriers dropout

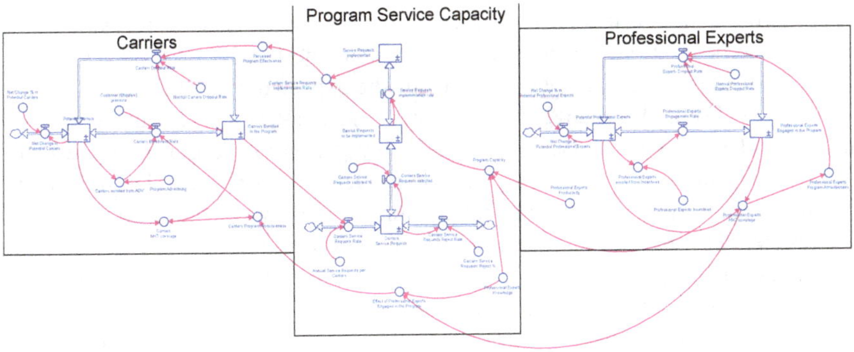

Fig. 5.2 A preliminary stock and flow structure of the EVE the environmental voluntary engagement program

rate. To overcome such a potential limit to growth (see B1 in Fig. 5.1) the EVE program coordinator may incentive professional experts to engage in the program.

5.5.4 *Post-COVID Implications for the Transport and Logistics Sector of France?*

The complexity of the supply chain in the transport and logistics sector is multi-dimensional. Stakeholders' perspectives, rising environmental concerns about CO_2 emissions, and unpredictable external shocks. The COVID-19 pandemic has added to the complexity of such supply chains. Decision makers and practitioners can overcome this challenge by utilizing professional expertise. Based on the findings of this study, transport operators can be engaged in environmental voluntary programs to build a durable competitive advantage, while complying with environmental policies, particularly in the post-COVID situation.

5.6 Conclusions

France, similarly to other European and non-European countries, launched an environmental voluntary program in the freight industry to improve transport operators' efficiency and to reduce CO_2 emissions. However, it has been demonstrated that these programs do not seem sufficient to drive, alone, a real change in the system. Several are the reasons for such a potential failure. Among the others, the difficulties to understand the multiple relationships between public and private actors involved in the program, the complexity of the transport and logistics sector, the uncertainly and hypercompetitive market segment may contribute to preventing the achievement of the desired project outcomes. To support the French EVE coordinator to design and assess effective policies to pursue the expected program goals, a System Dynamics (SD) model is suggested. As such, the analysis of the context in which all stakeholders involved operate and the investigation of the transport operators' sub-groups profiles will offer the ground to model the processes underlying actor's program engagement and the implementation of transport operator efficiency and environmental initiatives. Furthermore, by designing appropriate key performance indicators (such as program attractiveness, transport operators' engagement, program effectiveness, program long-term sustainability), the SD model will support the EVE coordinator to test alternative policies and to assess the success of the program. As the research project is in the early stage, the chapter outlines the research path and the preliminary stock-and-flow structure. In the next steps of this research, the preliminary SD model will be used in a group-model building process setting to adapt the model to project' participants' viewpoints.

References

Abernethy, M. A., & Chua, W. F. (1996). A field study of control redesign: The impact of institutional processes on strategic choice. *Contemporary Accounting Research, 13*(2), 569–606.

Aktar, M. A., Alam, M. M., & Al-Amin, A. Q. (2021). Global economic crisis, energy use, CO2 emissions, and policy roadmap amid COVID-19. *Sustainable Production and Consumption, 26*, 770–781. https://doi.org/10.1016/j.spc.2020.12.029

Bass, F. M. (1969). A new product growth model for consumer durables. *Management Science, 15*, 215–227.

Bivona, E., & Ceresia, F. (2008). Building long-term manufacturer-retailer relationships through strategic human resource management policies: A system dynamics approach. In H. Qudrat-Ullah, J. Spector, & P. Davidsen (Eds.), *Complex decision making: Understanding complex systems*. Springer.

Bivona, E., Ceresia, F., & Tumminello, G. (2019). Overcoming managers' myopic decisions in a waste collection company: Lessons from a system dynamics-based research. *Journal of Modelling in Management, 14*(4), 1023–1041. https://doi.org/10.1108/JM2-01-2019-0028

Bivona, E., & Montemaggiore, G. B. (2010). Understanding short- and long-term implications of 'myopic' fleet maintenance policies: A system dynamics application to a city bus company. *System Dynamics Review, 26*(3), 195–215.

Boiral, O. (2006). La certification ISO 14001: une perspective néo institutionnelle. *Management International, 10*(3), 67–79.

Bynum, C., Sze, C., Kearns, D., Polovick, B., & Simon, K. (2018). An examination of a voluntary policy model to effect behavioral change and influence interactions and decision making in the freight sector. *Transportation Research Part D, 61*, 19–32.

Casey, T. R., & Töyli, J. (2012). Dynamics of two-sided platform success and failure: An analysis of public wireless local area access. *Technovation, 32*, 703–716.

Castka, P., & Balzarova, M. (2008). ISO 26000 and supply chains on the diffusion of the social responsibility standard. *International Journal of Production Economics, 111*, 274–286.

Chu, S. H., Yang, H., Lee, M., & Park, S. (2017). The impact of institutional pressures on green supply chain management and firm performance: Top management roles and social capital. *Sustainability*, 9(5), 764. https://doi.org/10.3390/su9050764

Climate Transparency. (2021). *Comparing G20 climate action towards net zero* (Climate Transparency Report). https://www.climate-transparency.org/wp-content/uploads/2021/10/CT2021-Highlights-Report.pdf

Dadhich, P., Genovese, A., Kumar, N., & Acquaye, A. (2015). Developing sustainable supply chain in the UK construction industry: A case study. *International Journal of Production Economics, 164*, 271–284.

Davydenko, I., Ehrler, V., de Ree, D., Lewis, A., & Tavasszy, L. (2014). Towards a global CO2 calculation standard for supply chains: Suggestions for methodological improvements. *Transportation Research Part d: Transport and Environment, 32*, 362–372.

Delmas, M. A., & Toffel, M. W. (2008). Organizational responses to environmental demands: Opening the black box. *Strategic Management Journal, 29*(10), 1027–1055.

DiMaggio, P. J., & Powell, W. (1983). The iron cage revisited institutional isomorphism and collective rationality in organizational fields. *American Sociological Review, 48*, 147–160.

Eisenmann, T., Parker, G., & Van Alstyne, M. (2011). Platform envelopment. *Strategic Management Journal, 32*(12), 1270–1285. https://doi.org/10.1002/smj.935

Forliano, C., Ferraris, A., Bivona, E., & Couturier, J. (2022). Pouring new wine into old bottles: A dynamic perspective of the interplay among environmental dynamism, capabilities development, and performance. *Journal of Business Research, 142*, 448–463. https://doi.org/10.1016/j.jbusres.2021.12.065

Forrester, J. W. (1961). *Industrial dynamics*. MIT Press, currently available from Pegasus Communications.

Glover J. L., Champion, D., Daniels, K. J., & Dainty, A. J. D. (2014). An institutional theory perspective on sustainable practices across the dairy supply chain. *International Journal of Production Economics, 152*, 102–111.

Harrison, G., Bivona, E., & Rossetti, R. (2020). Editorial: Special issue on simulation in transportation. *Journal of Simulation, 14*(4), 239–241. https://doi.org/10.1080/17477778.2020.1829514

IPCC. (2018). Summary for policymakers. In V. Masson-Delmotte, P. Zhai, H.-O. Pörtner, D. Roberts, J. Skea, P. R. Shukla, A. Pirani, W. Moufouma-Okia, C. Péan, R. Pidcock, S. Connors, J. B. R. Matthews, Y. Chen, X. Zhou, M. I. Gomis, E. Lonnoy, T. Maycock, M. Tignor, & T. Waterfield (Eds.), *Global Warming of 1.5°C: An IPCC Special Report on the impacts of global warming of 1.5°C above pre-industrial levels and related global greenhouse gas emission pathways, in the context of strengthening the global response to the threat of climate change* (pp. 1–24). Cambridge University Press.

Kazancoglu, Y., Ozbiltekin-Pala, M., Sezer, M. D., Ekren, B. Y., & Kumar, V. (2022). Assessing the impact of COVID-19 on sustainable food supply chains. *Sustainability, 14*(1), 143. https://doi.org/10.3390/su14010143

Kumar, P., & Singh, R. K. (2021). Strategic framework for developing resilience in agri-food supply chains during COVID 19 pandemic. *International Journal of Logistics Research and Applications*. https://doi.org/10.1080/13675567.2021.1908524

Kunc, M. H., & Morecroft, J. D. (2010). Managerial decision making and firm performance under a resource-based paradigm. *Strategic Management Journal, 31*(11), 1164–1182.

Le Quéré, C., Korsbakken, J. I., Wilson, C., Tosun, J., Andrew, R., Andres, R. J., Canadell, J. G., Jordan, A., Peters, G. P., & van Vuuren, D. P. (2019). Drivers of declining CO 2 emissions in 18 developed economies. *Nature Climate Change, 9*(3). https://doi.org/10.1038/s41558-019-0419-7

Loannou, I., & Serafeim, G. (2017). *The consequences of mandatory corporate sustainability reporting* (Harvard Business School Research Working Paper, n°11–100).

Luna-Reyes, L. F., & Andersen, D. L. (2003). Collecting and analyzing qualitative data for system dynamics: Methods and models. *System Dynamics Review, 19*, 271–296.

Mendy Bilek, G., Fabbes-Coste, N., Kacioui-Maurin, E., Lazzeri, J., & Roussat, C., (2017). Obligation d'affichage des informations CO2 Transport: de la règlementation aux pratiques? Logistique & Management, Issue 2.

Meyer, J. W., & Rowan, B. (1977). Institutional organizations: Formal structure as myth and ceremony. *American Journal of Sociology, 83*(2), 340–363.

Minx, J. C., Lamb, W. F., Andrew, R. M., Canadell, J. G., Crippa, M., Döbbeling, N., Forster, P. M., Guizzardi, D., Olivier, J., Peters, G. P., Pongratz, J., Reisinger, A., Rigby, M., Saunois, M., Smith, S. J., Solazzo, E., & Tian, H. (2021). A comprehensive and synthetic dataset for global, regional, and national greenhouse gas emissions by sector 1970–2018 with an extension to 2019. *Earth System Science Data, 13*, 5213–5252. https://doi.org/10.5194/essd-13-5213-2021

OECD/ITF. (2015). *ITF Transport Outlook 2015*. OECD Publishing.

Prajogo, D., & Olhager J. (2012). Supply chain integration and performance: The effects of long-term relationships, information technology and sharing, and logistics integration. *International Journal of Production Economics, 135*, 514–522.

Rahmandad, H. (2015). Connecting strategy and system dynamics: An example and lessons learned. *System Dynamics Review, 31*(3), 149–172.

Repenning, N. P. (2000). A dynamic model of resource allocation in multi-project research and development systems. *System Dynamics Review, 16*(3), 173–212.

Richardson, G. P., & Pugh, A. L. I. I. I. (1981). *Introduction to system dynamics modeling with DYNAMO*. Productivity Press.

Roberts, N. H., Andersen, D. F., Deal, R. M., Grant, M. S., & Shaffer, W. A. (1983). *Introduction to computer simulation: The system dynamics modeling approach*. Addison-Wesley.

Romme, A. G. L., Zollo, M., & Berendsy, P. (2010). Dynamic capabilities, deliberate learning and environmental dynamism: A simulation model. *Industrial and Corporate Change, 19*(4), 1271–1299. https://doi.org/10.1093/icc/dtq031

Ruutu, S., Casey, T., & Kotovirta, V. (2017). Development and competition of digital service platforms: A system dynamics approach. *Technological Forecasting and Social Change, 117*, 119–130.

Sarkis, J., Gonzalez-Torre, P., & Diaz, B. (2010). Stakeholder pressure and the adoption of environmental practices: The mediating effect of training. *Journal of Operations Management, 28*(2), 163–176.

Sayed, M., Hendry L. C., & Zorzini Bell, M. (2017). Institutional complexity and sustainable supply chain management practices. *Supply Chain Management, 22*(6), 542–563.

Scholtens, B., & Kleinsmann, R. (2011). Incentives for subcontractors to adopt CO2 emissions reporting and reduction techniques. *Energy Policy, 39*(3), 1877–1883.

Sterman, J. (2000). *Business dynamics: Systems thinking and modeling for a complex world*. Irwin/McGraw-Hill.

Sterman, J. D. (1989). Modeling managerial behavior: Misperceptions of feedback in a dynamic decision making experiment. *Management Science, 35*(3), 321–339.

Tan, K. C., & Blanco, E. E. (2009, June). *System dynamics modeling of the SmartWay transport partnership*. MIT Engineering Systems Division & CESUN. Presented at the Second International Symposium on Engineering Systems at the MIT Center for Transportation and Logistics, Cambridge, Massachusetts.

Tate, W. L., Ellram, L. M., & Dooley, K. J. (2012). Environmental purchasing and supplier management (EPSM): Theory and practice. *Journal of Purchasing and Supply Management, 18*(3), 173–188.

Täuscher, K., & Laudien, S. M. (2018). Understanding platform business models: A mixed methods study of marketplaces. *European Management Journal, 36*, 319–329.

Vennix, J. A. M. (1996). *Group Model Building: Facilitating Team Learning Using System Dynamics*. Wiley.

William, F. L., Grubb, M., Diluiso, F., & Minx, J. C. (2022). Countries with sustained greenhouse gas emissions reductions: An analysis of trends and progress by sector. *Climate Policy, 22*(1), 1–17. https://doi.org/10.1080/14693062.2021.1990831

Wolmarans, P., Hyland, E., Atherton, S., Bovet, D., Bryan, J., Cheng, A. (2014). *Sustainability strategies addressing supply-chain air emissions*. National Cooperative Freight Research Program and the Transportation Research Board of National Academies, 28.

Zhu Q. (2016). Institutional pressures and support from industrial zones for motivating sustainable production among Chinese manufacturers. *International Journal of Production Economics, 181*, 402–409.

Zhu, Q., & Sarkis, J. (2007). The moderating effects of institutional pressures on emergent green supply chain practices and performance. *International Journal of Production Research, 25*(18–19), 4333–4355.

Chapter 6
Waste Processing Scenarios to Support Sustainable Environmental Development Using System Dynamics

Erma Suryani, Rully Agus Hendrawan, Ulfa Emi Rahmawati, and Dinda Meidianti Kusuma Putri

Abstract Waste cannot be separated from people's daily activities, resulting in a lot of waste accumulation in various areas, especially in densely populated areas. Waste management is now a common problem. The waste management paradigm must be based on the concept of waste management that supports the principles of sustainable environmental development. System dynamic modeling is used to formulate policies related to waste management because of its ability to solve problems systematically and consider the internal and external factors that affect sustainable environmental development. Several alternative scenarios were developed to support environmentally sustainable development. The scenarios developed include (1) recycling processed food waste by turning it into compost with the help of larvae; (2) burning waste that is safe and environmentally friendly with the help of an incinerator (burning) for waste that is not suitable for use in the composting process, and (3) conducting socialization and community training to make compost and burn environmentally friendly waste to prevent flood disasters and dangers in the health sector of the community. This scenario can be a recommendation for the government and other related parties in formulating strategies and policies related to waste management to reduce waste accumulation that supports sustainable environmental development.

Keywords Waste management · Model · Simulation · System dynamics · Environmentally friendly · Sustainable development · Scenarios · Prevent flood disasters · Strategy · Policy formulation · Problem-solving · Recycling processed food · Socialization · Community training · Health sector · Community · Policy recommendation · Causal loop diagram · Model validation · Garbage buildup · Post-Covid

---

E. Suryani (✉) · R. A. Hendrawan · U. E. Rahmawati · D. M. K. Putri
Department of Information Systems, Institut Teknologi Sepuluh Nopember (ITS), Surabaya, Indonesia
e-mail: erma.suryani@gmail.com

H. Qudrat-Ullah (ed.), *Understanding the Dynamics of New Normal for Supply Chains*, Understanding Complex Systems,
https://doi.org/10.1007/978-3-031-07333-5_6

6.1 Introduction

This section demonstrates the research background on the problem of the accumulation of household waste and medical waste due to the Covid-19 pandemic in Keputih Tegal Village, Surabaya City. Waste cannot be separated from people's daily activities, resulting in a lot of waste accumulation in various areas, especially in densely populated areas. The Covid-19 pandemic has caused the amount of waste to increase throughout the country due to the lockdown and social distancing policies (Sarkodie & Owusu, 2020). Waste management is now a common problem in Indonesia, such as in Keputih Tegal Village, Surabaya City. The waste generated by the people in Surabaya reaches 2,800 tons per day with the largest proportion (43.5%) of the waste generated coming from households (Maharrani & Syaifudin, 2020). The increase in household waste is a result of work from home and distance school policies (Febianto, 2021). The volume of single-use packaging waste is increasing due to the use of ready-to-eat food delivery services, take-away food services, and grocery packages from supermarkets (Popfalushi & Lviv, 2021). In addition, the increasing habit of people shopping online during the Covid-19 pandemic has increased the amount of waste. Types of plastic waste, cardboard, Styrofoam, and waste from materials commonly used to wrap other packages increased by around 27–36% (Febianto, 2021). Meanwhile, the average amount of mask waste reaches 863.15 kg per month (Surabaya City Government, 2021).

Waste that is not managed properly can negatively impact health and the environment because food waste can become a nest for animals such as flies, mice, foxes, and cockroaches (Ecube Labs, 2016; Mulyanti & Fachrurozi, 2016). Several diseases due to garbage accumulation include diarrhea, bubonic plague, dysentery, salmonella which causes typhoid fever, enteric fever, gastroenteritis, and other major diseases (Ecube Labs, 2016). In addition, people in the waste disposal area often burn the garbage, which causes air pollution. The problem of waste has not become a popular issue in Indonesia. In fact, according to a report by The Economic Intelligence Unit, Indonesia is the second-largest waste disposal country in the world (Folia, 2019). To overcome the waste problem in Surabaya, good cooperation with several parties, including the government, is needed. The waste management paradigm must be based on the concept of waste management that supports the principles of sustainable environmental development (Wan et al., 2019).

Based on these problems, a waste treatment solution is required to assist local governments in obtaining waste management policies to reduce their accumulation, so that they can support sustainable environmental development. The system dynamics simulation model is used to formulate policies related to waste management (Churchman, 1968; Sterman, *Business dynamics: System thinking and modeling for a complex world*, 2000). System dynamics modeling is used to explicitly solve problems systematically and model non-linear behavior and dynamic interactions (feedback) between interrelated factors with scenarios to reduce food waste accumulation (Ford, 1999; Walters et al., 2016). Proper waste management requires appropriate and affordable technology, accepted by the community, and environmentally friendly. System dynamics is a computer-based approach that helps predict system behavior and relate the dependent and independent variables (Popli et al., 2017) in the system. The problem of public resistance to new waste processing technology can delay the development of waste processing capacity (Sancheta et al., 2021).

Several alternative scenarios were developed to reduce waste accumulation, so that it can support sustainable and environmentally friendly development, including (1) recycling processed food waste by turning it into compost with the help of larvae; (2) burning waste that is safe and environmentally friendly with the help of an incinerator (burning) for waste that is not suitable for use in the composting process, and (3) conducting socialization and community training to make compost and burn environmentally friendly waste to prevent flood disasters and dangers in the health sector in the community. The alternative scenario can be a recommendation for the government and other related parties in formulating strategies and policies related to waste management to reduce its accumulation and support sustainable environmental development.

6.2 Literature Review

This section contains a literature review on the definition of waste and the impact of Covid-19 on garbage buildup. This literature review is useful to support research on good waste management.

6.2.1 Waste

Waste is considered worthless, unnecessary, or thrown away. Humans produce too much waste and cannot handle it sustainably (Dictionary, 2021). Waste that is not biodegradable and cannot be recycled properly has polluted the ocean and the environment. Waste affects the environment in many ways: its contribution to a worsening climate crisis, its negative impact on wildlife and the natural environment, and its harm to public health. Human health is threatened due to landfill waste (Downs & Acevedo, 2019). The types of waste include household waste, liquid and solid, hazardous waste, medical/clinical waste, electrical waste, recyclable waste, construction, and demolition debris, and green/organic waste (4waste, 2016; Steve, 2020). Household waste is any waste material that comes from the household. These households include private residences, hotels, dormitories, messes, campsites, picnic areas, and recreational areas (Insider, 2021). The amount of household waste generated by households can vary depending on income and lifestyle. There are several ways to reduce household waste, including donating clothes, eliminating single-use plastic, and choosing sustainable items, e.g., bamboo toothbrushes over plastic toothbrushes (Resource Center, 2019).

6.2.2 *Impact of Covid-19 on Garbage Buildup*

The Covid-19 pandemic has caused the amount of waste to increase throughout the country due to the lockdown and social distancing policies. The intensification of single-use products and panic buying have increased production and consumption, thus thwarting efforts to reduce plastic pollution (Sarkodie & Owusu, 2020). During the pandemic, personal protective equipment (PPE) has increased plastic pollution (Adyel, 2020). In addition, medical waste is increasingly being generated by health care facilities such as hospitals, doctor's practices, dental practices, veterinary hospitals/clinics, and medical research facilities and laboratories (EPA, 2020). Failure to manage waste generated from health facilities and households can increase the spread of Covid-19 through secondary transmission. The potential for rampant exhaust, open burning, and incineration can affect air quality and health outcomes due to exposure to toxins (WHO, 2020). Improper management of medical waste can potentially expose patients, health workers, and waste managers to injury, infection, toxic consequences, and air pollution (Mihai, 2020). For example, discarded needles can expose waste workers to needle stick injuries and potential infection when containers break in garbage trucks or needles are sent incorrectly to recycling facilities (EPA, 2020).

6.3 Methodology

This section discusses the method to be used in the research, such as discussing the theory and stages of the system dynamics model development. Waste that is not managed properly can have a negative impact on health and the environment (Ecube Labs, 2016; Mulyanti & Fachrurozi, 2016). The waste management paradigm must be based on the concept of waste management that supports the principles of sustainable environmental development (Wan et al., 2019). The system dynamics simulation model can be used as a method to formulate policies related to waste management (Churchman, 1968; Sterman, 2000). System dynamics modeling is used to explicitly solve problems systematically and model non-linear behavior and dynamic interactions (feedback) between interrelated factors with scenarios to reduce food waste accumulation (Ford, 1999; Walters et al., 2016).

The development of a system dynamics model is carried out through several stages (Sterman, 2000) which include: (1) system understanding; (2) Causal loop diagram (CLD) and Stock and Flow Diagram (SFD) development; (3) model formulation; (4) testing using behavior validity tests which are carried out by comparing the average (error rate) and amplitude variation (error variance) (Barlas, 1989; Qudrat-Ullah, 2012). Scenario Development in this case the scenario of waste management to reduce the accumulation of waste and support sustainable environmental development.

The model is defined as a system representation consisting of several variables that are interconnected with the specified dynamic problems (Richardson & Pugh,

1986). Simulation is the process of designing a model of a real system and conducting experiments with the model to understand the behavior of the system or evaluate various strategies for the operation of the system (Tasrif, 2004). System dynamics is a method to improve understanding in complex systems (Sterman, 2000). System dynamics can model non-linear behavior as well as dynamic interactions (feedback) between interrelated factors and can be handled easily by performing action scenarios or system changes (Walters et al., 2016). System dynamics, which was first introduced by Jay W. Forrester at the Massachusetts Institute of Technology (MIT), is a method of solving complex problems that arise due to causal tendencies of various variables in the system (Forrester, 1999).

6.4 Result and Discussion

This section explains the results and discussion of research on waste management using the system dynamics simulation model.

6.4.1 System Understanding and Causal Loop Diagram (CLD) Development

An understanding of the system is required as a basis for developing a causal loop diagram (CLD). CLD is required to describe the interrelationships of several variables that affect the number of waste heaps as well as several alternative waste management strategies to reduce waste accumulation that supports sustainable environmental development, as shown in Fig. 6.1.

The Covid-19 pandemic has caused the amount of waste to increase due to the lockdown and social distancing policies (Sarkodie & Owusu, 2020). The largest generation of waste in Surabaya comes from households (Maharrani & Syaifudin, 2020). The increase in household waste is a result of the work from home and distance school policies (Febianto, 2021), and the size of the population and per capita income. In addition, medical waste is increasingly being produced by health care facilities that are increasingly crowded due to the Covid-19 pandemic (EPA, 2020). The increasing population in Surabaya is also one of the causes of the increase in household waste, while the birth rate and death rate influence the population itself. The birth rate will increase the population, while the death rate will reduce the population.

Several alternative scenarios have been developed to reduce the accumulation of waste, including (1) recycling processed waste by turning it into compost, before being composted, waste is sorted and larvae assist the composting process; (2) burning waste that is safe and environmentally friendly with the help of an incinerator (burning) for waste that is not suitable for use in the composting process, the burned waste also goes through a first sorting stage; (3) conducting socialization

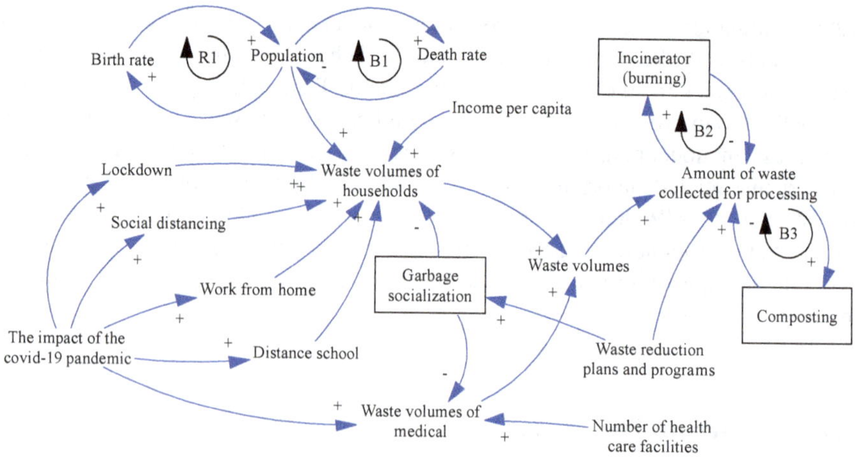

Fig. 6.1 CLD of waste management for waste reduction

to the community to make compost and burn environmentally friendly waste, this socialization aims to educate the public.

6.4.2 Stock and Flow Diagram (SFD) Development

The conceptual model described through the CLD is then converted into a dynamic system model described through the SFD which contains levels, rates, auxiliary, source, and sinks (Sterman, *Business dynamics: System thinking and modeling for a complex world*, 2000). SFD of the waste accumulation model can be seen in Fig. 6.2.

Model formulation of waste accumulation model can be seen in Eqs. 6.1–6.4:

$$Population\,(t+1) = Initial\ population\,(t0) + \int_{t0}^{t} (Population\,increase\,(t) - Population\,decrease\,(t))\,dt \tag{6.1}$$

$$Waste\,volumes\,of\,households\,(t+1) = Initial\,Waste\,volumes\,of\,households\,(t0) + \int_{t0}^{t} (Addition\,of\,household\,waste\,(t) - Reduction\,of\,household\,waste\,(t))\,dt \tag{6.2}$$

$$Waste\,volumes\,of\,medical\,(t+1) = Initial\,waste\,volumes\,of\,medical\,(t0) + \int_{t0}^{t} (Addition\,of\,medical\,waste\,(t) - Reduction\,of\,medical\,waste\,(t))\,dt \tag{6.3}$$

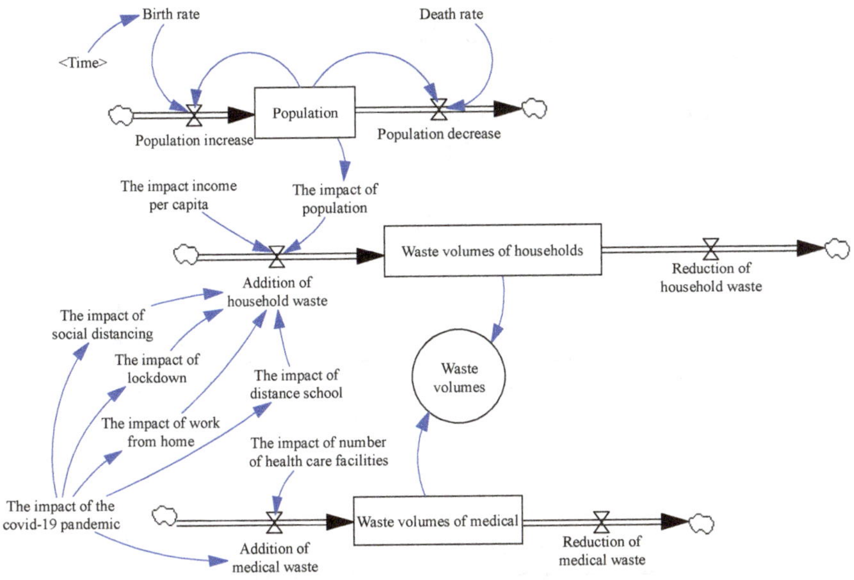

Fig. 6.2 SFD of waste volumes

$$\begin{aligned} Waste\ volumes\,(t) &= Waste\ volumes\ of\ households\,(t) \\ &\quad + Waste\ volumes\ of\ medical\,(t) \end{aligned} \tag{6.4}$$

The simulation result of the Surabaya City population sub-model is shown in Fig. 6.3. As we can see from Fig. 6.3, the population in Surabaya City tends to fluctuate from 2008 to 2020 with an average increase of 0.2%. The average population in the period 2008–2020 was around 2,949,339 people.

Meanwhile, the simulation result of waste volumes of the household's sub-model is shown in Fig. 6.4. Waste volume of households increased from 2008 to 2020, with an average increase of 0.009%. The average waste volume of households in 2008 to 2020 is around 444,571.85 Tons.

The simulation result of the waste volume of the medical sub-model is shown in Fig. 6.5. The waste volume of medical increased from 2008 to 2020 with an average increase of 0.02%. The average waste volume of medical from 2008 to 2020 is around 214,540.92 Tons.

The simulation result of the total waste volumes sub-model is shown in Fig. 6.6. Waste volume increased from 2008 to 2020 with an average increase of 0.01%. The average waste volume from 2008 to 2020 was around 659,112.77 Tons.

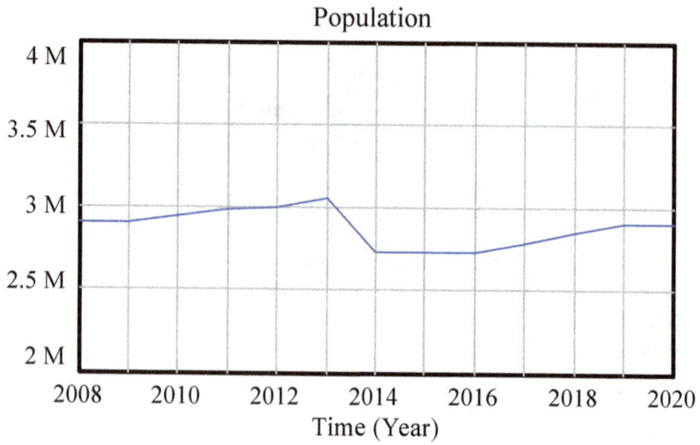

Fig. 6.3 Simulation result of Surabaya population

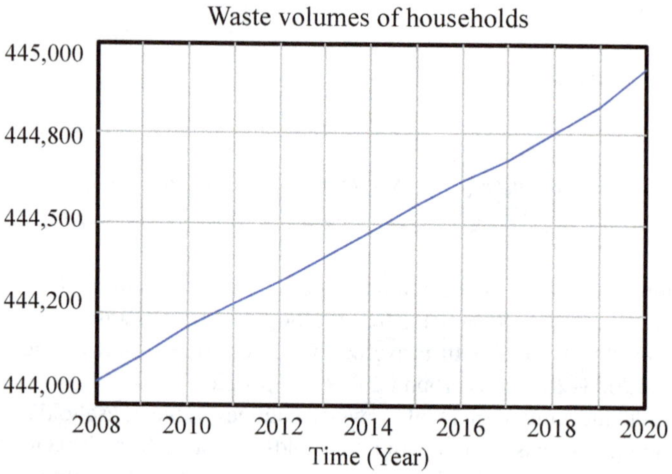

Fig. 6.4 Simulation result of waste volumes of households

6.4.3 Model Validation

Model validation can be done by comparing the average or error rate and variations in amplitude or error variance (Barlas, 1989; Qudrat-Ullah, 2012). The model is categorized as valid if the error rate is 5% and the error variance is 30%. The model validation calculations are found in Eqs. 6.5 and 6.6.

1. Average Comparison Test or *Error Rate*

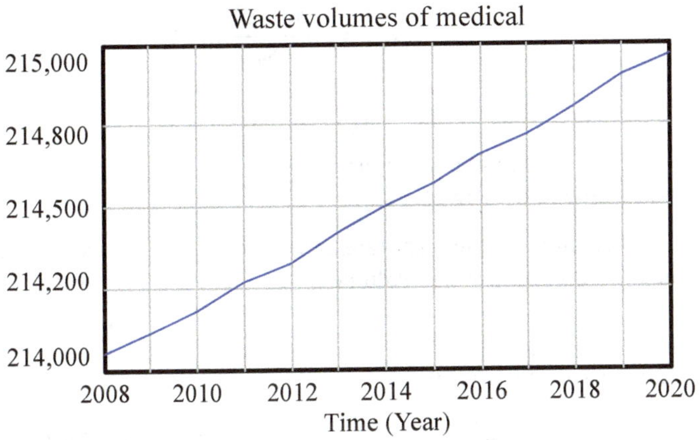

Fig. 6.5 Simulation result of waste volumes of medical

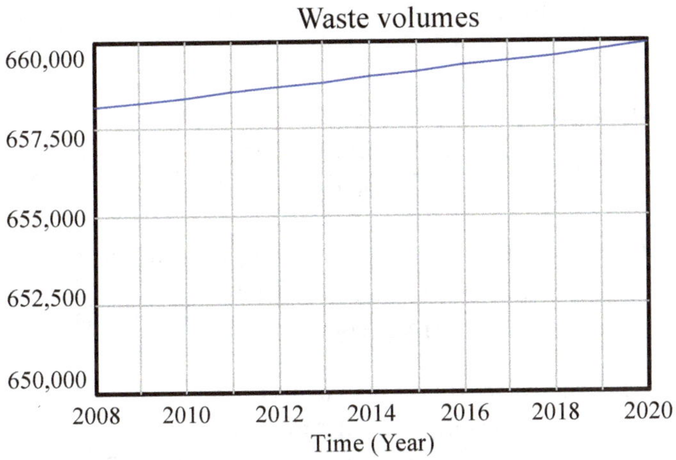

Fig. 6.6 Simulation result of total waste volumes

$$Error\ rate = \left| \frac{\overline{S} - \overline{A}}{\overline{A}} \right| \times 100\% \quad (6.5)$$

where:

$\overline{S}$ The average rate of simulation
$\overline{A}$ The average rate of data

2. Comparative Test of Amplitude Variation or *Error Variance*

$$Error\ variance = \left| \frac{S_s - S_a}{S_a} \right| \times 100\% \qquad (6.6)$$

where:

S_S Standard deviation of simulation
S_a Standard deviation of Data

The results of calculating the error rate and error variance in the population, the waste volume of households, waste volume of medical, and total waste volume are as follows:

$$Error\ rate\ of\ population = \left| \frac{2{,}879{,}688.462 - 2{,}949{,}339}{2{,}949{,}339} \right| \times 100\% = 2.36\%$$

$$Error\ variance\ of\ population = \left| \frac{107{,}647.3388 - 105{,}533.1912}{105{,}533.1912} \right| \times 100\% = 2.00\%$$

$$Error\ rate\ of\ waste\ volumes\ of\ households = \left| \frac{444{,}482.5385 - 444{,}571.8462}{444{,}571.8462} \right| \times 100\% = 0.02\%$$

$$Error\ variance\ of\ waste\ volumes\ of\ households = \left| \frac{273.7135411 - 291.4223299}{291.4223299} \right| \times 100\% = 6.08\%$$

$$Error\ rate\ of\ waste\ volumes\ of\ medical = \left| \frac{214{,}496.6154 - 214{,}540.9231}{214{,}540.9231} \right| \times 100\% = 0.02\%$$

$$Error\ variance\ of\ waste\ volumes\ of\ medical = \left| \frac{301.2315108 - 296.7542366}{296.7542366} \right| \times 100\% = 1.51\%$$

$$Error\ rate\ of\ waste\ volumes = \left| \frac{658{,}979.2308 - 659{,}112.7692}{659{,}112.7692} \right| \times 100\% = 0.02\%$$

$$Error\ variance\ of\ waste\ volumes = \left| \frac{574.6848635 - 585.0728094}{585.0728094} \right| \times 100\% = 1.78\%$$

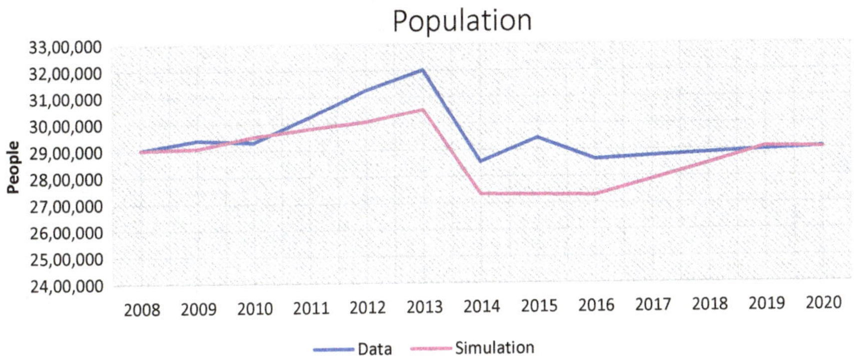

Fig. 6.7 The comparison of data and population submodel

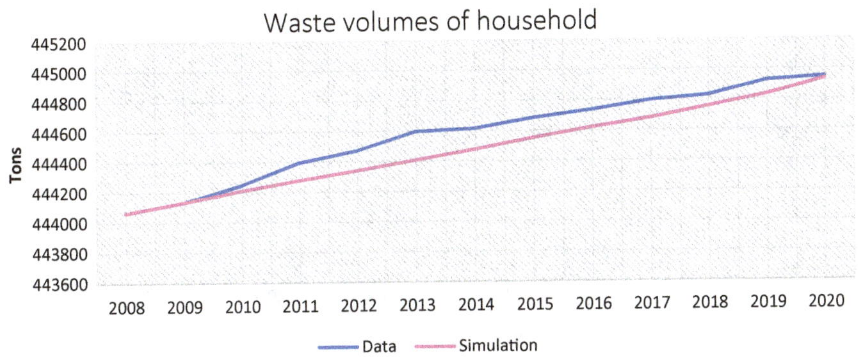

Fig. 6.8 The comparison of data and waste volumes of households submodel

From the error rate and error variance test results, all error rates are $\leq 5\%$ and error variance are $\leq 30\%$, thus indicating that the model is valid. The comparison graph of the simulation results with the data can be seen in Figs. 6.7, 6.8, 6.9, and 6.10.

6.4.4 Scenario

Based on the causal loop diagram (CLD) and stock and flow diagram (SFD) that have been made, several alternative scenarios are obtained to reduce waste accumulation so that it can support environmentally sustainable development, especially in Keputih Tegal Village, Surabaya City. The waste reduction scenario model can be seen in Fig. 6.11. This scenario was developed by considering the following factors:

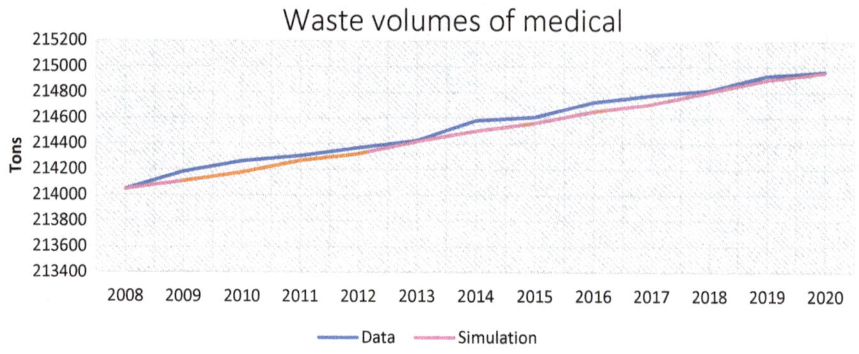

Fig. 6.9 The comparison of data and waste volumes of medical submodel

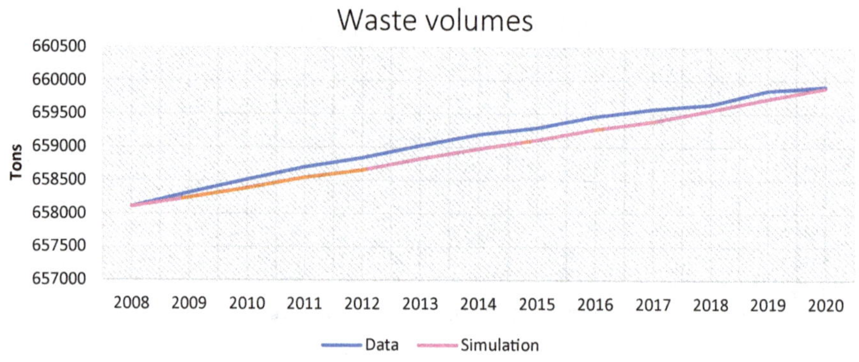

Fig. 6.10 The comparison of data and total waste volumes submodel

(a) Recycle processed waste by turning it into compost. Compost can be made from organic waste, both from animals and plants, so it is very environmentally friendly. Compost is an alternative for processing organic waste into a new form that humans and the environment can reuse. This will have a better and more natural impact. In addition, the process of making compost can be said to be quite easy. Composting can be done independently on a household scale. This is what makes non-chemical fertilizers begin to be widely discussed amid the issue of global warming and environmental damage. By re-managing waste into compost, it can increase the income of the people of Keputih Tegal Village, Surabaya City. Here are some of the benefits of compost:

- Increasing soil fertility
- Improving soil characteristics and structure
- Increasing microbial activity in the soil
- Increasing groundwater absorption
- Improving crop quality

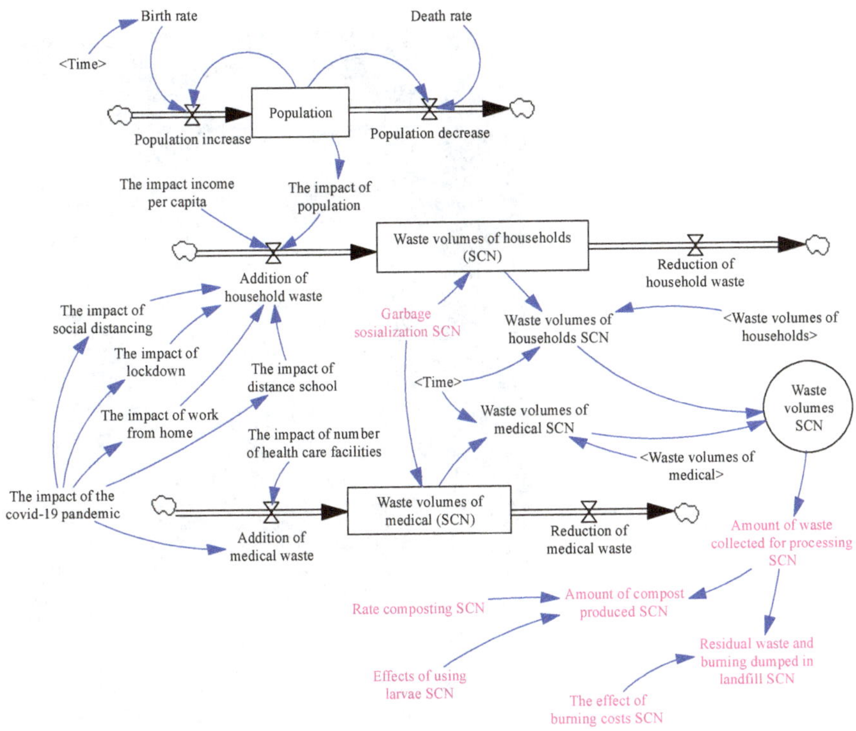

Fig. 6.11 Scenario model for reducing waste accumulation

- Providing nutrients to the soil
- Making the soil looser and less hard
- Suppressing the growth of plant diseases
- Providing vitamins and hormones needed by plants.

(b) Burning waste in a safe and environmentally friendly manner

One alternative technology has been developed to deal with waste problems on a micro to macro scale. This technology is known as an incinerator or a waste-burning device. Incinerator technology optimally burns waste with complete combustion until the waste becomes environmentally friendly ash. Incinerators have been widely used in various cities in Indonesia, but the incinerators used are still not optimal. It is not only because they are expensive but also not able to answer all problems related to waste and the environment. Generally, these tools are imported from abroad, costing billions of rupiah and requiring trained operators and technicians. This external incinerator in operation is quite expensive because destroying waste requires a large amount of fuel and electricity continuously. In addition, tool components are not easily available in the domestic market, so it is quite troublesome when damage and maintenance occurs. The example of an incinerator is shown in Fig. 6.12.

Fig. 6.12 Domestic Incinerator

Incinerator technology is designed to have some ease to operate. Some of the advantages of incinerators are:

- They do not require a large space
- They can burn dry waste to wet waste
- The destruction of the combustion system reaches temperatures above 900 °C
- They work effectively without additional fuel
- The level of pollution is low. In operations in several places, it is proven that the smoke from the combustion that comes out of the chimney is almost invisible and does not emit a disturbing odor
- The temperature of the hot air exhaust in the chimney is constantly controlled
- The temperature of the outside walls remains cool, equal to the temperature of the outside air
- They need easy and inexpensive maintenance
- Burning ash can be processed into a variety of building material products.

(c) Conducting socialization and community training

Socialization and training to the community of Keputih Tegal Village, Surabaya City was carried out as initiation and education of the actions that the community needed to take regarding the solutions offered. Through socialization, people are given insight that the waste they usually produce and collect

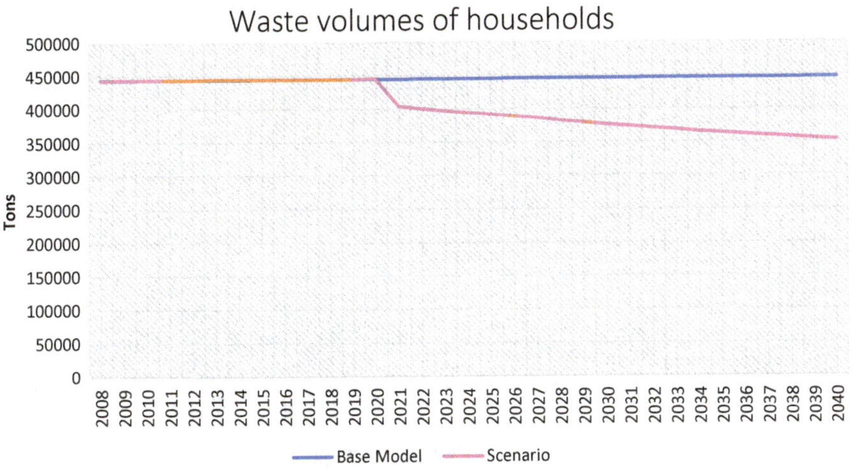

Fig. 6.13 The waste volume of households before (base model) and after the scenario

can be of high value rather than just being thrown into the trash. Meanwhile, the training will give examples of how to make compost and burn environmentally friendly waste. In addition, the purpose of socialization in waste management is to prevent and anticipate flood disasters and health hazards in the community.

The simulation results of the waste volumes of the household's model scenario before (base model) and after the scenario can be seen in Fig. 6.13.

The scenario simulation results show that the waste volume of households has decreased from an average of 445,315 Tons/Year to an average of 402,992 Tons/Year. The waste volume of households decreased by around 10.50%. The waste volume of medical before (base model) and after the scenario can be seen in Fig. 6.14.

The scenario simulation results show that the waste volume of medical has decreased from an average of 215,232 Tons/Year to an average of 194,716 Tons/Year. The waste volume of medical decreased by about 10.54%. The comparison of waste volume before (base model) and after the scenario can be seen in Fig. 6.15.

The scenario simulation results show that waste volumes have decreased from an average of 660,668 Tons/Year to an average of 597,709 Tons/Year. Waste volume decreased by about 10.53%. It can be concluded that this scenario can be a recommendation for the government and other related parties in formulating strategies and policies related to waste management to reduce its accumulation and support sustainable environmental development during the Covid-19 pandemic.

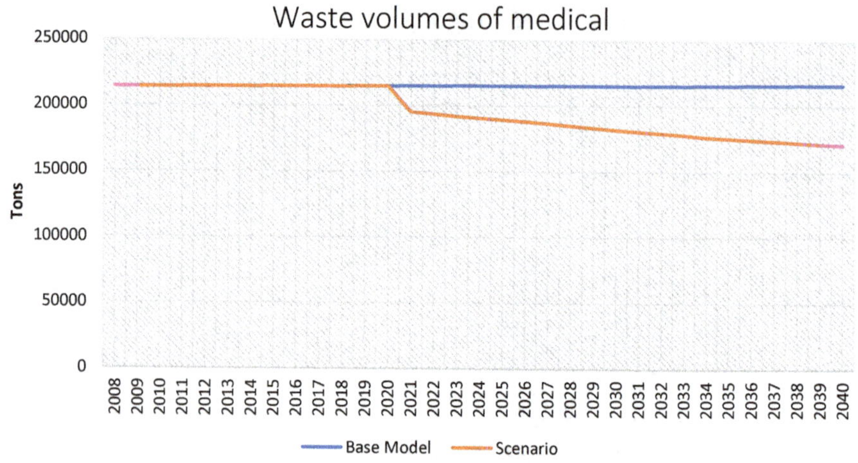

Fig. 6.14 The waste volume of medical before (base model) and after the scenario

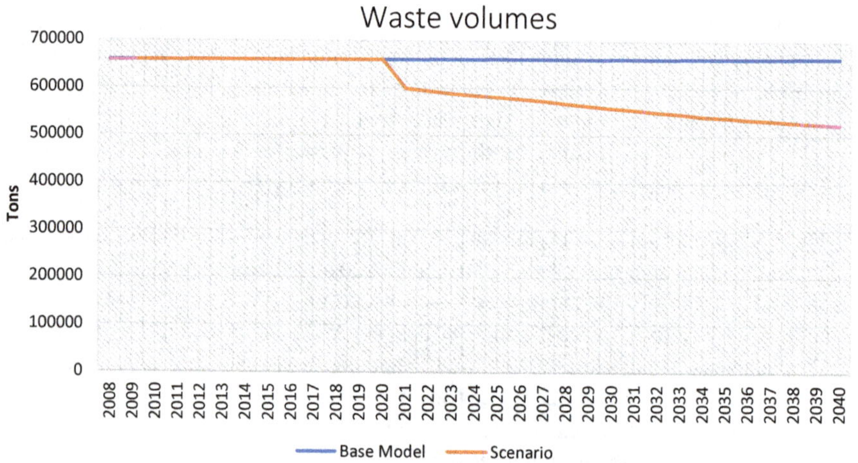

Fig. 6.15 Total waste volume before (base model) and after the scenario

6.5 Implications for Post-Covid Waste Management Supply Chains

Based on the research results, it can be concluded that post-Covid waste management is required to improve the waste management supply chain efficiency. Three alternative scenarios were developed to reduce waste accumulation, such as (1) recycling processed waste by turning it into compost; (2) burning waste that is safe and environmentally friendly with the help of an incinerator (burning), and (3) conducting socialization to the community to make compost and burn environmentally friendly waste. They are considered quite optimal in increasing the efficiency of the post-Covid waste management supply chain because they can reduce waste accumulation maximally and adequately, in terms of proper and efficient use of resources. The results show that the waste volume gradually decreased by about 10.53%.

6.6 Conclusion and Further Research

This research is designed to reduce waste accumulation to support environmentally sustainable development through the development of waste management models for waste reduction. The accumulation of waste referred to here is the accumulation of waste volume of households and waste volume of medical. This research was conducted by developing the best scenario using dynamic system modeling to accommodate all elements of the problem in determining strategies related to reducing waste accumulation.

Several factors that affect the accumulation of waste during the Covid-19 pandemic include population size, per capita income, lockdown policies, social distancing, work from home, and distance school. These factors are significant variables that affect the accumulation of waste in Keputih Tegal Village, Surabaya City. Scenario development was done by changing the structure of the validated model. Several scenarios developed include (1) recycling the processed waste by turning it into compost with the help of larvae; (2) burning waste that is safe and environmentally friendly with the help of an incinerator (burning) for waste that is not suitable for use in the composting process, and (3) conducting socialization and community training to make compost and burn environmentally friendly waste to prevent flood disasters and dangers in the health sector in the community. The scenario simulation results show that: (1) waste volume of households decreased by around 10.50%; (2) waste volume of medical decreased by about 10.54%; and (3) total waste volume decreased by about 10.53%. It can be concluded that this scenario can be a recommendation for the government and other related parties in formulating strategies and policies related to waste management to reduce its accumulation and support sustainable environmental development during the Covid-19 pandemic.

References

4waste. (2016). *There are 5 types of waste, do you know them all?* Retrieved October 10, 2021, from https://4waste.com.au/rubbish-removal/5-types-waste-know/

Adyel, T. M. (2020). Accumulation of plastic waste during COVID-19. *Science, 369*(6509), 1314–1315.

Barlas, Y. (1989). Multiple tests for validation of system dynamics type of simulation models. *European Journal of Operational Research, 42*, 59–87.

Churchman, C. W. (1968). *The systems approach*. Dell Publishing Co., Inc.

Dictionary. (2021). *Dictionary*. Retrieved October 12, 2021, from https://www.dictionary.com/browse/trash
Downs, A., & Acevedo, R. (2019). *How our trash impacts the environment*. Retrieved October 12, 2021, from https://www.earthday.org/how-our-trash-impacts-the-environment/
Ecube Labs. (2016). *Overflowing garbage bins: 5 impacts on health and environment, and how to prevent*. Retrieved September 4, 2021, from https://www.ecubelabs.com/overflowing-garbage-bins-5-impacts-on-health-and-environment-and-how-to-prevent/
EPA. (2020). *EPA United States Environmental Protection Agency*. Retrieved October 12, 2021, from https://www.epa.gov/rcra/medical-waste
Febianto, R. (2021). *Jumlah sampah melonjak selama pandemi COVID-19, ini beberapa penyebabnya*. Retrieved September 4, 2021, from https://www.liputan6.com/health/read/4483554/jumlah-sampah-melonjak-selama-pandemi-covid-19-ini-beberapa-penyebabnya
Folia, R. (2019). *Eva Bachtiar, Pahlawan Sampah Makanan di Surabaya*. Retrieved September 3, 2021, from https://www.idntimes.com/life/women/rosa-folia/eva-bachtiar-pahlawan-sampah-makanan-di-surabaya-1/2
Ford, A. (1999). *Modeling the environment: An introduction to system dynamics models of environmental systems*. Island Press.
Forrester, J. W. (1999). *System dynamics: The foundation under system thinking*. Sloan School of Management MIT.
Insider, L. (2021). *Household waste definition*. Retrieved October 9, 2021, from https://www.lawinsider.com/dictionary/household-waste
Maharrani, A., & Syaifudin, N. (2020). *Ada asa pengelolaan sampah di Surabaya*. Retrieved September 4, 2021, from https://lokadata.id/artikel/ada-asa-kelola-sampah-di-surabaya
Mihai, F. C. (2020). Assessment of COVID-19 waste flows during the emergency state in Romania and related public health and environmental concerns. *International Journal of Environmental Research and Public Health, 17*(15), 5439.
Mulyanti, K., & Fachrurozi, A. (2016). Analisis sikap dan perilaku masyarakat terhadap pelaksanaan program bank sampah [studi kasus masyarakat Kelurahan Bahagia Bekasi Utara]. *Jurnal Ilmiah Ekonomi Manajemen dan Kewirausahaan "Optimal", 10*(2), 185–198.
Popfalushi, D., & Lviv, Z. W. (2021). *The impact of the COVID-19 pandemic on the waste management system in Ukraine and what problems in this area have been identified*. Retrieved September 4, 2021, from https://zerowasteeurope.eu/2021/04/the-impact-of-the-covid-19-pandemic-on-the-waste-management-system-in-ukraine/
Popli, K., Sudibya, G. L., & Kim, S. (2017). A review of solid waste management using system dynamics modeling. *Journal of Environmental Science International, 26*(10), 1185–1200.
Qudrat-ullah, H. (2012). On the validation of system dynamics type simulation model. *Telecommunication Systems, 51*(2–3), 159–166.
Resource Center. (2019). *What is household waste?* Retrieved October 7, 2021, from https://www.buschsystems.com/resource-center/knowledgeBase/glossary/what-is-household-waste
Richardson, G. P., & Pugh, A. L. (1986). *Introduction to system dynamics modelling with dynamo*. The MIT Press.
Sancheta, L. d., Chaves, G. d., & Siman, R. R. (2021). The use of system dynamics on urban solid waste management: a literature analysis. *Gestão & Produção*, 1–18.
Sarkodie, S. A., & Owusu, P. A. (2020). Impact of COVID-19 pandemic on waste management. *Environment, Development and Sustainability*, 1–10.
Sterman, J. D. (2000). *Business dynamics: System thinking and modeling for a complex world*. Jeffrey J. Shelstad.
Steve. (2020). *7 Different types of rubbish you need to dispose of*. Retrieved October 11, 2021, from https://www.stevesrubbishremovals.com.au/7-different-types-of-rubbish-you-need-to-dispose-of/
Surabaya City Government. (2021). *Sebulan bisa capai 863 kilogram, begini cara pemkot tangani sampah rumah tangga masker*. Retrieved September 4, 2021, from https://www.surabaya.go.id/id/berita/61997/sebulan-bisa-capai-863-kilogram

Tasrif, M. (2004). *Model simulasi untuk analisis kebijakan: pendekatan metodologi system dynamics*. Institut Teknologi Bandung.

Walters, J. P., Archer, D. W., Sassenrath, G. F., Hendrickson, J. R., Hanson, J. D., Halloran, J. M., Vadas, P., & Alarcon, V. J. (2016). Exploring agricultural production systems and their fundamental components with system dynamics modelling. *Ecological Modelling, 333*, 51–65.

Wan, C., Shen, G. Q., & Choi, S. (2019). Waste management strategies for sustainable development. In W. Leal Filho (Ed.), *Encyclopedia of Sustainability in Higher Education* (pp. 1–9). Springer International Publishing.

WHO. (2020). *Water, sanitation, hygiene, and waste management for SARS-CoV-2, the virus that causes COVID-19*. Jenewa: World Health Organization and the United Nations Children's Fund (UNICEF).

Chapter 7
Multi-criteria Analysis of Disruption Risks for Supply Chains Due to Pandemics

J. Martino Neto and Valerio Antonio Pamplona Salomon

Abstract Coronavirus Disease 2019 (COVID-19) affected global economics and society, unprecedentedly. Supply chains, linking customers, manufacturers, and suppliers, are more susceptible to disruption risks when facing pandemics, like COVID-19. As matter of fact, there is an emerging literature on supply chain management (SCM) and COVID-19. This chapter explores how supply chain managers may evaluate supply chain risks due to pandemics. Managers may analyze alternatives to mitigate the situation. The purpose of this chapter is to present a mathematical model for assessing disruption risks in supply chains affected by pandemics. A multi-criteria decision analysis (MCDA) model is developed from the consolidated literature of SCM. Analytic Hierarchy Process (AHP) and Technique of Order Preference by Similarity to Ideal Solution (TOPSIS), two leading MCDA methods were combined in the development of the assessment model. The model is tested with the case study of a multinational automotive company that operates in both efficient and responsive supply chains. For efficient supply chains, the model resulted in a focus on capacity management, demand planning, and sales forecasting, to avoid risks disruptions. For responsive supply chains, the focus shall move to operations management.

Keywords Supply chain management · Coronavirus Disease 2019 · Risk management · Efficient supply chain · Responsive supply chain · Multi-criteria decision analysis · Analytic hierarchy process · Technique of Order Preference · Ideal Solution · Automotive industry · Latin America · Model · Demand planning · TOPSIS · Case study · Suppliers · Customers · Agile supply chains

7.1 Introduction

Supply chains will no longer be the same, post-pandemic as the Coronavirus Disease 2019 (COVID-19). The need to adapt to the new scenario added local and global

---

J. M. Neto · V. A. P. Salomon (✉)
Department of Production, Sao Paulo State University, Av. Dr. Ariberto P. Cunha 333, Guaratingueta, SP 12516-410, Brazil
e-mail: salomon@feg.unesp.br; valerio.salomon@unesp.br

H. Qudrat-Ullah (ed.), *Understanding the Dynamics of New Normal for Supply Chains*, Understanding Complex Systems,
https://doi.org/10.1007/978-3-031-07333-5_7

transformations in societies, impacting everyone's life. Then, companies, corporations, organizations, and entire supply chains also had to review their processes and the ways of operating them. Chopra (2019) categorized supply chains, regarding processes, as efficient supply chains or responsive supply chains. Efficient supply chains seek to supply demands at the lowest possible costs, with lower margins, lower service levels, higher utilization of assets, and cheaper transportation modes. On the other hand, responsive supply chains meet flexible order requirements, shorter lead times, and wider product mixes (Anparasan & Lejeune, 2018; Holweg, 2005). In practice, supply chain management (SCM) analyses and defines proper mixes of efficiency and responsiveness. Supply chains must perform well according to financial indicators, as well as for customers' satisfaction, with sustainability. However, in a post-pandemic scenario, are the previous mixes of efficiency and responsiveness still proper?

Regional epidemics have been discussed as a potential source of risk to supply chains (Anparasan & Lejeune, 2018; Farooq et al., 2021). However, companies may not be prepared for global pandemics (Ivanov, 2020). Risk management in supply chains is a set of preventive and responsive actions to decrease damages and losses to the whole supply chain (Mailena et al., 2021). The purpose of this chapter is to propose a model to risk assessment of efficient and responsive supply chains. For that, a literature review is carried out, evaluating the effects of COVID-19. A multicriteria decision analysis (MCDA) model (Ishizaka & Nemery, 2013) is developed to assess benefits, costs, opportunities, and risks. The MCDA model was obtained with a hybrid application of the Analytic Hierarchy Process (AHP) and the Technique of Order Preference by Similarity to Ideal Solution (TOPSIS).

In addition to this Introduction, this chapter has five more sections: Sect. 7.2 presents the Literature Review; Sect. 7.3, Methodology; Sect. 7.4, Case Study; Sect. 7.5, Results and Discussion; and Sect. 7.6, Conclusions.

7.2 Literature Review

Supply chain management (SCM) is all the activities of procuring materials and services, converting unfinished goods into final products, connecting suppliers and customers globally (Boström et al., 2015; Scuotto et al., 2017). Supply chains were firstly categorized by Fisher (1997) as "physically efficient" or "market-responsive". Chopra (2019) corroborates this categorization, as presented in Table 7.1.

Several important factors should be taken into account in the process of building an adaptive supply chain. Fluctuating demand, market uncertainty, and the emergence of new technologies explain the need for more flexible and agile supply chains (Hajiagha et al., 2021; Zidi et al., 2021). However, in a post-pandemic scenario, is the previous mix of efficiency and responsiveness still proper? How to assess the benefits, costs, opportunities, and risks involved?

Agile supply chains were a trend in supply chain management, in pre-COVID-19 times (Azevedo & Sousa, 2000). It is possible to define agility as a business-wide

Table 7.1 Comparison of efficient and responsive supply chains (Chopra, 2019; Fisher, 1997)

	Efficient supply chain	Responsive supply chain
Primary goal	Supply to demand at the lowest costs	Respond quickly to demand
Product design	Increase performance with low costs	Allow product differentiation
Pricing	Lower margins	Higher margins
Manufacturing	High equipment utilization	Flexible due to demand uncertainty
Inventory	Reduced levels to lower costs	Buffer stocks of parts & finished goods
Lead time	Reduced, but not increasing costs	Aggressively reduced, even increasing costs
Supplier selection	Based on costs, then quality	Based on flexibility, reliability, and speed, then quality

capability that embraces organizational structures, information systems, logistics processes, and in particular, mindsets (Christopher, 2000; Ismail & Sharifi, 2006).

Therefore, the agile supply chain regards adaptability and flexibility, because of its continual, dynamic, and quick response to customers' changing needs in competitive environments (Gunasekaran et al., 2008; Lin et al., 2005). The focus is on speed, responsiveness, cost efficiency, and an increase in the productivity of goods and services (Gawade, 2021). All these features can be identified in responsive supply chains.

Due to COVID-19, SCM has been deeply transformed from 2020. Pandemics are extraordinarily disruptive, in many ways, including all dimensions of forwarding and reverse logistics. Their long duration generates problems with negative impacts on multiple levels of societies (Ivanov & Dolgui, 2019). There is an ambiguity of when or how the pandemic can be considered under control (Gölgeci & Kuivalainen, 2020). COVID-19 brought to supply chains more uncertainties than ever before (Durmaz et al., 2021): Broken international supply links, steep price increases from suppliers, disruption of transportation, insufficient raw materials for many industrial sectors, financial vulnerabilities, delays in delivery due to transportation issues during lockdowns, and risks of spreading infection from contact deliveries (Dhama et al., 2020; Sarkis et al., 2020; Yu & Aviso, 2020). All these facts have severely obstructed supply chains operation across the world (Antony et al., 2021; Jabbour et al., 2020). COVID-19 is considered a major disruptive event of this decade, raising unforeseen social-economic implications worldwide (Leite et al., 2020).

The impact of COVID-19 pandemics, and the measures for its control, have changed the normal, challenging supply chain ecosystems, networks, flows, and individual firms on unprecedented scales, under severe uncertainty (Ivanov & Dolgui, 2020; Sodhi et al., 2021). With the shutting down of manufacturing units, during several months, supply chain managers have tried to develop alternatives for supply chain operations (Birkel & Hartmann, 2019; Calatayud et al., 2019), need to reorganize themselves to ensure continuity of operations and future availability of products (Flynn et al., 2021; Wang et al., 2020).

Table 7.2 Causes and sources of disruption risks for supply chain management

Source	Causes	References
Customers	Changes in customer behaviour, demand variations, facility closures and local lockdown	Göçer (2021), Kumar et al. (2021)
Manufacturers	Disinformation from suppliers or customers, efficiency or productivity decreases, facility closures, and material issues	Belhadi et al. (2021), Birkel and Hartmann (2019), Calatayud et al. (2019)
Suppliers	Auditing delays, disinformation from customers, global transportation collapses, logistics and material issues, and networks redesign	Bag et al. (2021), Hajiagha et al. (2021), Ivanov (2020), Nassereddine et al. (2021), Velayutham et al. (2021), Zidi et al. (2021)

Risk can be considered a major factor in making different decisions (Mokhtarzadeh et al., 2020a, 2020b). According to Jabbarzadeh et al. (2020), there are two main types of risks entailed in the supply chain network context:

- Disruption risks: Low frequent disruptions, usually caused by disasters or large-scale threats inflicting considerable damages on the whole supply chain.
- Operational risks: Ingrained uncertainties, caused by mundane disturbances with high frequency in nature and small interruptions.

Disruption risks are the focus of post-pandemic analyses. Table 7.2 presents causes of disruption risks grouped by customer, manufacturer, and supplier.

Risk management provided assessment and risk mitigation in a case of SCM in the construction industry (Banaitiene & Banaitis, 2012). Risk management has become extremely important in post-pandemic scenarios, with rapidly changing environments, bringing new and more adaptive approaches (Bakos & Dumitras, 2021). Bocanet et al. (2021) studied the effects of pandemics in business environments and the society, analysing employment growth rates. Kumar et al. (2021) identified and analyzed risk mitigation strategies for supply chains of perishable food. El Baz and Ruel (2021) surveyed practices of risk management to mitigate disruption impacts on supply chains in the COVID-19 outbreak. Alkahtani et al. (2021) proposed a non-linear model to provide economic benefit to a supply chain with high demand fluctuation due to COVID-19. Qazi et al. (2021) explored the efficacy of early warning systems in predicting pandemics. Bag et al. (2021) proposed the use of big data to help to restore strength to supply chains. Di Francesco et al. (2021) developed a model using contract-based mechanisms considering the risks of demand uncertainty, supply disruption, and random yield. Friday et al. (2021) proposed a collaborative approach to maintain optimal inventory, mitigating disruption risks during pandemics. Li et al. (2021) summarized the current finances of supply chains and their trends. Ivanov (2020) presents results of simulated impacts

Table 7.3 Research on risk management applied to supply chain management considering pandemics effects

Reference	Research objective
Ahlqvist et al. (2021)	Propose the concept of supply chain risk management governance
Alkahtani et al. (2021)	Develop a non-linear supply chain management model to deal with the different situations under variable demand
Bag et al. (2021)	Use of big data to restore strength to the supply chain
Bocanet et al. (2021)	Study effects of pandemics in the business of United Arab Emirates
Di Francesco et al. (2021)	Provide guidelines for managers on supply chain's risks influences
El Baz and Ruel (2021)	Study the integration of information and material flows
Friday et al. (2021)	Review literature on collaborations in healthcare supply chains and reinforce resilience against disruptions during pandemics
Ivanov (2020)	Simulate COVID-19 impacts in global supply chains
Kumar et al. (2021)	Discusses all mitigation strategies concerning the socioeconomic contingencies originating from COVID-19
Li et al. (2021)	Summarize the situation of supply chain finance and identify future trends
Qazi et al. (2021)	Explore the efficacy of disasters and epidemics-based risk ratings in predicting the country-level exposure to COVID-19

from COVID-19 to global supply chains. Ahlqvist et al. (2021) developed the concept of "supply chain risk governance" that embraces various types of supply chain actors. Table 7.3 presents a summary of previous researches on risk management applied to SCM considering pandemics effects.

Several works have already addressed disruption risks in the supply chain caused by pandemics. This chapter innovates with the proposal of multi-criteria decision analysis (MCDA). In Sect. 7.3, two MCDA methods are presented: The Analytic Hierarchy Process (AHP) and the Technique of Order Preference by Similarity to Ideal Solution (TOPSIS). In Sect. 7.4, a hybrid model of MCDA with the application of AHP and TOPSIS methods is presented.

7.3 Methods

This section introduces concepts and steps for the AHP application and the TOPSIS application. AHP and TOPSIS are two leading methods for MCDA applied in the SCM (Khan et al., 2018). AHP is presented in Sect. 7.3.1 and TOPSIS in Sect. 7.3.2.

7.3.1 Analytic Hierarchy Process

Developed from Saaty (1974) and proposed in Saaty (1980), AHP is one of the most applied MCDA methods. There are AHP applications in diverse areas, like chemical engineering, computer science, ecology, energy sector, health sector, higher education sector, manufacturing, mathematical advances, supply chain management, and logistics (Emrouznejad & Marra, 2017).

The first step for the AHP application is to model the decision problem in a hierarchy. Decision objective is on the top, n criteria are in the middle of the hierarchy, and m alternatives are in the bottom. Figure 7.1 presents a hierarchy model for $n = 3$ and $m = 4$.

The meaning of a hierarchical model is that the elements in a lower level need to be assessed regarding the elements in the higher level. Then, criteria shall be pairwise compared regarding the decision objective. The Saaty Scale, more commonly referred to as the Fundamental Scale of Absolute Numbers (Saaty, 2013), is used for pairwise comparisons. The Saaty Scale is a linear 1–9 scale, with 1 for "equal importance", 3 for "weak importance of one over another", 5 for "strong importance", 7 for "very strong importance", and 9 for "absolute importance". Intermediate values, as 2, 4, 6, and 8, and even rational numbers may be used, if needed.

Weights for the criteria are obtained normalizing the right eigenvector w of the pairwise comparison matrix A, as in Eq. (7.1), where λ_{max} is its maximum eigenvalue.

$$Aw = \lambda_{max} w \tag{7.1}$$

Consistency checking is one of the great advantages of AHP against other MCDA methods. A consistent pairwise matrix A satisfies $a_{ij} = a_{ik}a_{kj}$, resulting in $\lambda_{max} = n$. Consistency index μ is a measure of the consistency of a pairwise matrix, as in Eq. (7.2), where n is the number of criteria.

$$\mu = \frac{\lambda_{max} - n}{n - 1} \tag{7.2}$$

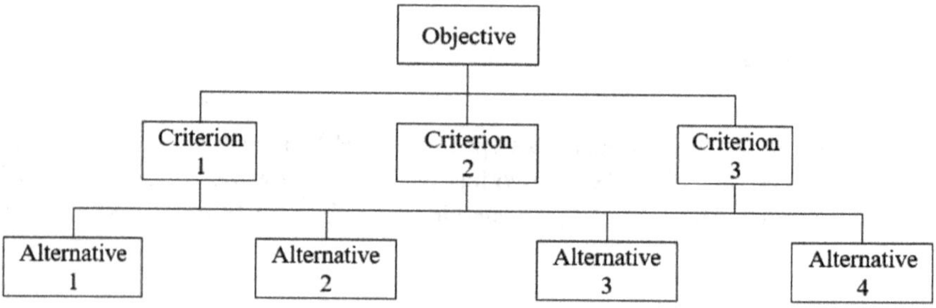

Fig. 7.1 Three-level hierarchy for a decision problem with three criteria and four alternatives

Consistency ratio CR is a better measure since it compares μ with a random index RI, computed by the Oak Ridge Laboratory, with more than 50,000 matrices (Saaty, 1980), as in Eq. (7.3).

$$CR = \frac{\mu}{RI} \tag{7.3}$$

Consistent matrices have $\lambda_{max} = n$, then $\mu = 0$ and $CR = 0$. Inconsistent matrices have at least one comparison, and its reciprocal, $a_{ij} \neq a_{ik}a_{kj}$, resulting $\lambda_{max} > n$. It is desirable that $CR \leq 0.1$, then A may be accepted, meaning "conformity with previous practice" or that decision-makers did not change their minds when fulfilling a pairwise comparison matrix.

Alternatives shall be assessed regarding each criterion, resulting in performance x_{ij} of Alternative i regarding Criterion j. The same procedure of pairwise comparisons may be used to obtain the preferences of the alternatives regarding each criterion. In this chapter, a hybrid AHP–TOPSIS model is proposed: AHP to weight the criteria; TOPSIS for alternatives assessment. Then, the next subsection presents TOPSIS concepts and procedures.

The main limitations of AHP are the independence and the number of alternatives and criteria. If some alternatives or criteria have dependency on each other, then the Analytic Network Process (ANP) would be a proper MCDA method (Saaty, 2009). If $n > 9$, then criteria must be aggregated or grouped as sub-criteria, which will lead to one more level in the hierarchy model. If $m > 9$, alternatives can be grouped, or absolute measurement may be applied (Saaty, 1986; Salomon, 2016). Overall performances of alternatives y_i are obtained weighting local performances of alternatives x_{ij} by weights of criteria w_j, as in Eq. (7.4).

$$y_i = \sum_{j=1}^{n} x_{ij} w_j \tag{7.4}$$

7.3.2 *Technique of Order Preference by Similarity to Ideal Solution*

TOPSIS assesses performances of alternatives through similarity with the ideal solution (Hwang & Yoon, 1981; Lombardi Netto et al., 2021). Table 7.4 can be used to determine performance values x_{ij} for alternatives i regarding criteria j.

Pairwise comparisons are not performed in TOPSIS. Components of decision matrix x_{ij} are obtained weighting the performance values v_{ij} by the weights of criteria w_j, as in Eq. (7.5):

$$x_i = v_{ij} w_j \tag{7.5}$$

Table 7.4 Level of performance

Level	Performance
Excellent	1.0
Very high	0.9
High	0.8
Medium	0.7
Low medium	0.6
Low	0.5
Very low	0.4

Besides its name referring to an ideal solution, TOPSIS also considers an anti-ideal solution, also referred to negative ideal solution. Negative ideal solution a_j^-, and positive ideal solution a_j^+, can be obtained as in Eqs. (7.6 and 7.7).

$$a_j^- = \min(x_{ij}) \tag{7.6}$$

$$a_j^+ = \max(x_{ij}) \tag{7.7}$$

Then, Euclidean distances to negative and positive solutions, respectively d^- and. d^+, can be obtained as in Eqs. (7.8 and 7.9).

$$d_i^- = \sqrt{\sum_{j=1}^{n} \left(x_{ij} - a_j^-\right)^2} \tag{7.8}$$

$$d_i^+ = \sqrt{\sum_{j=1}^{n} \left(x_{ij} - a_j^+\right)^2} \tag{7.9}$$

Finally, closeness coefficients ci are obtained as in Eq. (7.10).

$$c_i = \frac{d_i^-}{d_i^+ + d_i^-} \tag{7.10}$$

When performances of Alternative i are closer to the positive solution and further to the negative solution, then $c_i > 0.5$.

7.4 Case Study

A multinational automotive manufacturing corporation operates two industrial plants in Brazil: Units 1 and 2. Both units assemble cars, delivered in efficient and responsive

Table 7.5 Experts in supply chain management

Expert	Occupation	Age	Experience in company
1	Director of supply chain	58 years old	32-year
2	Supply chain manager (Unit 1)	52 years old	35-year
3	Supply chain manager (Unit 2)	44 years old	18-year
4	Buyer (Unit 1)	46 years old	22-year
5	Buyer (Unit 2)	41 years old	17-year
6	Operations manager (Unit 1)	43 years old	14-year
7	Operations manager (Unit 2)	49 years old	12-year

chains. In Unit 1, located in the Brazilian state of Parana, the best-seller car has an efficient supply chain, since this is a popular car. In the same plant, more expensive models are delivered in a responsive supply chain.

Unit 2 is located in the interior of Rio de Janeiro state. The best-seller car is a sport utility vehicle. Curiously, sales of this model are divided into a responsive supply chain, for the regular market, and an efficient supply chain, for the special market of disabled people, referred to as PCD (*Pessoas com deficiência*), in Brazil.

Table 7.5 presents a group of experts in SCM from the company, selected to provide data for AHP–TOPSIS application on disruption risk assessment for supply chain due to pandemics impacts.

The literature reviewed in Sect. 7.2 was presented to the group of experts in a first video-conference session. This meeting resulted in the hierarchy model presented in Fig. 7.2, as a consensus by experts.

The criteria in the middle of Fig. 7.2's hierarchy came majorly from Table 7.1's first column. Experts also proposed another criterion: Transportation. Therefore, this is the set of criteria to assess causes for disruption risks at the bottom of the hierarchy, (Consumer Behavior, Demand Variation, Efficiency & Productivity Decreases, and Material Issues), all taken from Table 7.2's middle column.

In the second video-conference session, experts provided, by consensus, pairwise comparisons among the criteria, regarding efficient supply chains (Table 7.6) and also regarding responsible supply chains (Table 7.7). Both comparison matrices can be accepted since they resulted in $CR \approx 0.033$ (Table 7.6) and $CR \approx 0.087$ (Table 7.7).

Inventory and Transportation are the heaviest criteria for efficient supply chains. Surprisingly, these criteria were judged more important than Pricing, even for supply chains with price as "a prime customer drive" (Chopra, 2019). This was due to pandemic effects, from experts' point of view.

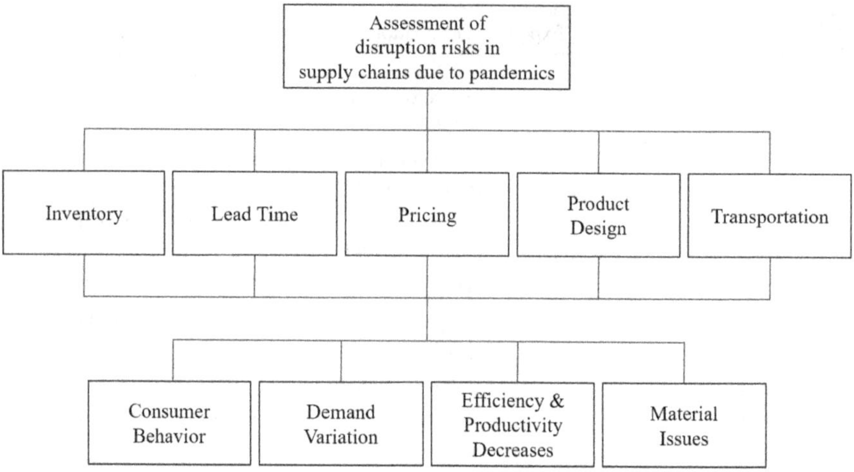

Fig. 7.2 Proposed hierarchy for assessment of disruption risks in supply chains due to pandemics

Table 7.6 Pairwise comparisons and weights of criteria for efficient supply chains

Criterion	INV	LDT	PRI	PRO	TPT	Weight (%)
Inventory (INV)	1	3	2	7	1	33
Lead time (LDT)	1/3	1	3	5	1/2	19
Pricing (PRI)	½	1/3	1	3	1/3	11
Product design (PRO)	1/7	1/5	1/3	1	1/7	4
Transportation (TPT)	1	2	3	7	1	33

Table 7.7 Pairwise comparisons and weights of criteria for responsive supply chains

Criterion	INV	LDT	PRI	PRO	TPT	Weight (%)
INV	1	7	7	5	3	49
LDT	1/7	1	1	1/5	1/7	4.5
PRI	1/7	1	1	1/5	1/7	4.5
PRO	1/5	5	5	1	1/3	14
TPT	1/3	7	7	3	1	28

Inventory is a far more important criterion for responsive supply chains, considering pandemics effects. As expected, Pricing decreased in importance. The importance of Lead Time also decreased, but Product Design's increased.

Still in the second video-conference session, experts also by consensus, assessed alternatives, regarding each criterion. Again, experts provided data for efficient supply chains (Table 7.8) and also for responsive supply chains (Table 7.9).

Table 7.8 Assessment of disruption risks causes for efficient supply chains

Disruption risks cause	INV	LDT	PRI	PRO	TPT
Consumer behavior	Very high	High	Medium	Very low	Medium
Demand variation	Very high	High	High	Very low	High
Efficiency and productivity decreases	High	High	Very high	Low	Low
Material issues	High	Very high	Low	Very low	High

Table 7.9 Assessment of disruption risks causes for responsive supply chains

Disruption risks cause	INV	LDT	PRI	PRO	TPT
Consumer behavior	Low	Low	Medium	Low	Medium
Demand variation	Very low	Low	Medium	Very low	Medium
Efficiency and productivity decreases	Very high	Very high	Very high	High	Medium
Material issues	High	Very high	Very high	Very high	High

Table 7.10 Decision matrix for efficient supply chains

Disruption risks cause	INV	LDT	PRI	PRO	TPT
Consumer behavior	0.297	0.152	0.077	0.016	0.231
Demand variation	0.297	0.152	0.088	0.016	0.264
Efficiency and productivity decreases	0.264	0.152	0.099	0.020	0.165
Material issues	0.264	0.171	0.055	0.016	0.264

According to experts, all alternatives have low or very low performance in Product Design. This means that all of them have a low chance to cause disruption risks impacting Product Design for efficient supply chains. Conversely, all alternatives have high or very high performance in Inventory. This means that they have a high chance to cause disruption risks impacting Inventory for efficient supply chains.

Performance of alternatives varied more regarding responsive supply chains, according to experts. For instance, Demand Variation kept very low performance in Product Design, but Material Issues increased from very low to very high.

The assessed values for alternatives presented in Tables 7.8 and 7.9 were associated with the performance values in Table 7.4 and weighted by criteria weights presented in Tables 7.6 and 7.7, respectively, resulting in Tables 7.10 and 7.11.

7.5 Results and Discussion

This section is divided into three subsections. Section 7.5.1 presents the Results; Sect. 7.5.2, managerial implications; and Sect. 7.5.3, post-pandemic implications.

Table 7.11 Decision matrix for responsive supply chains

Disruption risks cause	INV	LDT	PRI	PRO	TPT
Consumer behavior	0.245	0.022	0.031	0.070	0.196
Demand variation	0.196	0.022	0.031	0.056	0.196
Efficiency and productivity decreases	0.441	0.040	0.040	0.112	0.196
Material issues	0.392	0.040	0.040	0.126	0.224

Table 7.12 Closeness coefficient for risk disruption causes

Disruption risks cause	Efficient supply chains	Responsive supply chains
Consumer behavior	0.635	0.198
Demand variation	0.831	0.000
Efficiency and productivity decreases	0.294	0.889
Material issues	0.646	0.811

7.5.1 Results

Table 7.12 presents the joint AHP–TOPSIS results for the efficient and responsive supply chains.

Demand Variation was considered the main cause of risk disruption for efficient supply chains. On the other hand, Demand Variation was not considered as a cause of disruption risk for responsive supply chains. These may result from the responsiveness of this kind of supply chain.

For responsive supply chains, Efficiency and Productivity Decreases and Material Issues have closeness coefficients higher than 0.8. Therefore, these are two major causes for disruption risks of supply chains.

7.5.2 Implications for Supply Chain Managers

The managerial implications of the case study are that, when facing pandemic effects, supply chain managers must:

- When looking for efficiency, focus on Demand Variation, Material Issues, and Consumer Behavior. That is, focus on capacity management, demand planning, and sales forecasting.
- When looking for responsiveness, focus on operations to increase or keep efficiency and productivity indicators.

It is not too much to reinforce that these results came from a case study. Specifically, a case from the automotive industry, primarily located in Latin America. Noticing that, readers may generalize the results for different industries or locations, *mutatis mutandis*.

7.5.3 Post-COVID-19 Implications for Supply Chains

Post-COVID-19 implications:

1. The pandemic has become more volatile to the market and will involve the reconfiguration of supply chains due to the unpredictability of demand and the increased uncertainty.
2. Although there is a lot of research about the impacts on supply chains, the strategies to adapt to the new scenario will demand resilience from their managers in their respective markets, in the search for solutions aligned with their competitive advantage.
3. It is possible to highlight the increase in e-commerce, accelerating growth forecasts in this area, and change in infrastructure and distribution models of goods and consumption as consequence.

7.6 Conclusions

This chapter presented a multi-criteria analysis model to assess risk disruptions in the supply chain due to pandemic effects. The model was applied in a case from the automotive industry. Both efficient and responsive supply chains were studied. Two methods of multi-criteria analysis were combined: Analytic Hierarchy Process (AHP) and Technique of Order Preference by Similarity to Ideal Solution (TOPSIS). The set of criteria was obtained from consolidated SCM Theory (Chopra, 2019) and the set of alternatives came from the literature review, presented in Sect. 7.2.

Experts from a multinational corporation with industrial plants located in Brazil provided data for AHP and TOPSIS applications. Different weights were proposed for efficient and responsive supply chains. As a consequence, Demand Variation was considered as the main cause of disruption risks for efficient supply chains. On the other hand, Efficiency and Productivity Decreases were considered as the main causes of disruption risks for responsive supply chains.

As these results came from the Latin American automotive industry, the first proposal for future researches is the application of the model in other industries, as the chemicals or services industry and in other locations, like Asia or Europe.

Another interesting theme for research is the application of different multi-criteria methods. With the Analytic Network Process, for instance, it will be possible to analyze the dependency or influence among the alternatives and the criteria. With

Fuzzy Theory it will be possible to incorporate uncertainty elements to the analysis of pandemic effects in supply chain management.

References

Ahlqvist, V., Norrman, A., & Jahre, M. (2021). Supply chain risk governance: Towards a conceptual multi-level framework. *Operations and Supply Chain Management: An International Journal, 13*, 382–395.

Alkahtani, M., Omair, M., Khalid, Q. S., Hussain, G., Ahmad, I., & Pruncu, C. (2021). A COVID-19 supply chain management strategy based on variable production under uncertain environment conditions. *International Journal of Environmental Research and Public Health, 18*, 1662.

Anparasan, A. A., & Lejeune, M. A. (2018). Data laboratory for supply chain response models during epidemic outbreaks. *Annals of Operations Research, 270*, 53–64.

Antony, J., Psomas, E., Garza-Reyes, J. A., & Hines, P. (2021). Practical implications and future research agenda of lean manufacturing: A systematic literature review. *Production Planning & Control, 32*, 889–925.

Azevedo, A., & Sousa, J. P. (2000). Agile supply-chain management: Challenges, requirements and solutions. *IFAC Proceedings Volumes, 33*, 917–922.

Bag, S., Dhamija, P., Luthra, S., & Huisingh, D. (2021). How big data analytics can help manufacturing companies strengthen supply chain resilience in the context of the COVID-19 pandemic. *The International Journal of Logistics Management*. https://doi.org/10.1108/IJLM-02-2021-0095

Bakos, L., & Dumitras, D. D. (2021). Decentralized enterprise risk management issues under rapidly changing environments. *Risks, 9*, 165.

Banaitiene, N., & Banaitis, A. (2012). Risk management in construction projects. In N. Banaitiene (Ed.), *Risk management: Current issues* (pp. 429–448). InTech Open.

Belhadi, A., Kamble, S., Jabbour, C. J. C., Gunasekaran, A., Ndubisi, N. O., & Venkatesh, M. (2021). Manufacturing and service supply chain resilience to the COVID-19 outbreak: Lessons learned from the automobile and airline industries. *Technological Forecasting and Social Change, 163*, 120447.

Birkel, H. S., & Hartmann, E. (2019). Impact of IoT challenges and risks for SCM. *Supply Chain Management: An International Journal, 24*, 39–61.

Bocanet, A., Alpenidze, O., & Badran, O. (2021). Business analysis in post-pandemic era. *Academy of Strategic Management Journal, 4*, 1–9.

Boström, M., Jönsson, A. M., Lockie, S., Mol, A. P., & Oosterveer, P. (2015). Sustainable and responsible supply chain governance: Challenges and opportunities. *Journal of Cleaner Production, 107*, 1–7.

Calatayud, A., Mangan, J., & Christopher, M. (2019). The self-thinking supply chain. *Supply Chain Management: An International Journal, 24*, 22–38.

Chopra, S. (2019). *Supply chain management: Strategy, planning, and operation* (7th edn.). Pearson.

Christopher, M. (2000). The agile supply chain: Competing in volatile markets. *Industrial Marketing Management, 29*, 37–44.

Dhama, K., Khan, S., Tiwari, R., Sircar, S., Bhat, S., Malik, Y. S., Singh, K. P., Chaicumpa, W., Bonilla-Aldana, D. K., & Rodriguez-Morales, A. J. (2020). Coronavirus disease 2019–COVID-19. *Clinical Microbiology Reviews, 33*(4), e00028–e120.

Di Francesco, R. M., Meena, P., & Tibrewala, R. (2021). Buyback and risk-sharing contracts to mitigate the supply and demand disruption risks. *European Journal of Industrial Engineering, 15*(4), 550–581.

Durmaz, A., Demir, H., & Sezen, B. (2021). The role of negative entropy within supply chain sustainability. *Sustainable Production and Consumption, 28*, 218–230.

El Baz, J., & Ruel, S. (2021). Can supply chain risk management practices mitigate the disruption impacts on supply chains' resilience and robustness? Evidence from an empirical survey in a COVID-19 outbreak era. *International Journal of Production Economics, 233*, 107972.
Emrouznejad, A., & Marra, M. (2017). The state of the art development of AHP (1979–2017): A literature review with a social network analysis. *International Journal of Production Research, 55*, 6653–6675.
Farooq, M. U., Hussain, A., Masood, T., & Habib, M. S. (2021). Supply chain operations management in pandemics: A state-of-the-art review inspired by COVID-19. *Sustainability, 13*, 2504.
Fisher, M. L. (1997). What is the right supply chain for your product? *Harvard Business Review, 75*, 105–117.
Flynn, B., Cantor, D., Pagell, M., Dooley, K. J., & Azadegan, A. (2021). From the editors: Introduction to managing supply chains beyond COVID-19–preparing for the next global mega-disruption. *Journal of Supply Chain Management, 57*, 3–6.
Friday, D., Savage, D. A., Melnyk, S. A., Harrison, N., Ryan, S., & Wechtler, H. (2021). A collaborative approach to maintaining optimal inventory and mitigating stockout risks during a pandemic: Capabilities for enabling health-care supply chain resilience. *European Journal of Industrial Engineering, 11*, 248–271.
Gawade, D. (2021). Agile supply chain in manufacturing and service industry: Bibliometric and content analysis. *Vision*. https://doi.org/10.1177/09722629211002000
Göçer, F. (2021). A novel interval value extension of picture fuzzy sets into group decision making: An approach to support supply chain sustainability in catastrophic disruptions. *IEEE Access, 9*, 117080–117096.
Gölgeci, I., & Kuivalainen, O. (2020). Does social capital matter for supply chain resilience? The role of absorptive capacity and marketing–supply chain management alignment. *Industrial Marketing Management, 84*, 63–74.
Gunasekaran, A., Lai, K., & Cheng, T. C. E. (2008). Responsive supply chain: A competitive strategy in a networked economy. *Omega—International Journal of Management Science, 36*, 549–564.
Hajiagha, S. H. R., Mahdiraji, H. A., Behnam, M., Nekoughadirli, B., & Joshi, R. (2021). A scenario-based robust time–cost tradeoff model to handle the effect of COVID-19 on supply chains project management. *Operations Management Research*. https://doi.org/10.1007/s12063-021-00195-y
Holweg, M. (2005). An investigation into supplier responsiveness: Empirical evidence from the automotive industry. *International Journal of Logistics Management, 16*, 96–119.
Hwang, C. L., & Yoon, K. (1981). *Multiple attribute decision making: Methods and applications: A state-of-the-art survey*. Springer.
Ishizaka, A., & Nemery, P. (2013). *Multi-criteria decision analysis: Methods and Software*. Wiley.
Ismail, H. S., & Sharifi, H. (2006). A balanced approach to building agile supply chains. *International Journal of Physical Distribution, 36*, 431–444.
Ivanov, D. (2020). Predicting the impacts of epidemic outbreaks on global supply chains: A simulation-based analysis on the coronavirus outbreak (COVID-19/SARS-CoV-2) case. *Transportation Research Part e: Logistics and Transportation Review, 136*, 101922.
Ivanov, D., & Dolgui, A. (2019). New disruption risk management perspectives in sup- ply chains: Digital twins, the ripple effect, and resileanness. *IFAC Papersonline, 52*, 337–342.
Ivanov, D., & Dolgui, A. (2020). OR-methods for coping with the ripple effect in supply chains during COVID-19 pandemic: Managerial insights and research implications. *International Journal of Production Economics, 232*, 107921.
Jabbarzadeh, A., Fahimnia, B., & Sabouhi, F. (2020). Resilient and sustainable supply chain design: Sustainability analysis under disruption risks. *International Journal of Production Research, 56*, 5945–5968.
Jabbour, A. B. L. S., Song, M., & Godinho Filho, M. (2020). Sustainability implications for operations management: Building the bridge through exemplar case studies. *Production Planning & Control, 31*, 841–844.

Khan, S. A., Chaabane, A., & Dweiri, F. T. (2018). Multi-criteria decision-making methods application in supply chain management: A systematic literature review. In V. Salomon (Ed.), *Multi-criteria methods and techniques applied to supply chain management* (pp. 3–31). InTech Open.

Kumar, A., Mangla, S. K., Kumar, P., & Song, M. (2021). Mitigate risks in perishable food supply chains: Learning from COVID-19. *Technological Forecasting and Social Change, 166*, 120643.

Leite, H., Lindsay, C., & Kumar, M. (2020). COVID-19 outbreak: Implications on health-care operations. *The TQM Journal, 33*, 247–256.

Li, J., Wang, Y., Feng, G., Wang, S., & Song, Y. (2021). Supply chain finance review: Current situation and future trend. *System Engineering Theory and Practice, 40*, 1977–1995.

Lin, C. H., Chiu, H., & Chu, P. Y. (2005). Agility index in the supply chain. *International Journal of Production Economics, 100*, 285–299.

Lombardi Netto, A., Salomon, V. A. P., & Ortiz-Barrios, M. A. (2021). Multi-criteria analysis of green bonds: Hybrid multi-method applications. *Sustainability, 13*, 10512.

Mailena, L., Indrawanto, C., & Astuti, E. P. (2021). Risk management of chilli supply chains using weighted failure mode effect analysis. *IOP Conference Series: Earth and Environmental Science, 782*, 022004.

Mokhtarzadeh, N. G., Mahdiraji, H. A., Jafari-Sadeghi, V., Soltani, A., & Kamardi, A. A. (2020a). A product-technology portfolio alignment approach for food industry: A multi-criteria decision making with z-numbers. *British Food Journal, 122*, 3947–3967.

Mokhtarzadeh, N. G., Mahdiraji, H. A., Jafarpanah, I., Jafari-Sadeghi, V., & Cardinali, S. (2020b). Investigating the impact of networking capability on firm innovation performance: Using the resource-action-performance framework. *Journal of Intellectual Capital, 21*, 1009–1034.

Nassereddine, H., Seo, K. W., Rybkowski, Z. K., Schranz, C., & Urban, H. (2021). Propositions for a resilient, post-COVID-19 future for the AEC industry. *Frontiers in Built Environment, 7*, 687021.

Qazi, A., Simsekler, M. C. E., & Akram, M. (2021). Efficacy of early warning systems in assessing country-level risk exposure to COVID-19. *Geomatics, Natural Hazards and Risk, 12*, 2352–2366.

Saaty, T. L. (1974). Measuring fuzziness of sets. *Journal of Cybernetics, 4*, 53–61.

Saaty, T. L. (1980). *The analytic hierarchy process*. McGraw-Hill.

Saaty, T. L. (1986). Absolute and relative measurement with the AHP: The most livable cities in the United States. *Socio-Economic Planning Sciences, 20*, 327–331.

Saaty, T. L. (2009). *Theory and applications of the analytic network process: Decision making with benefits, opportunities, costs, and risks*. RWS.

Saaty, T. L. (2013). *Mathematical principles of decision making*. RWS.

Salomon, V. A. P. (2016). Absolute measurement and ideal synthesis in the AHP. *International Journal of the Analytic Hierarchy Process, 3*, 538–545.

Sarkis, J., Cohen, M. J., Dewick, P., & Schröder, P. (2020). A brave new world: Lessons from the COVID-19 pandemic for transitioning to sustainable supply and production. *Resources, Conservation, and Recycling, 28*, 104894.

Scuotto, V., Caputo, F., Villasalero, M., & Del Giudice, M. (2017). A multiple buyer–supplier relationship in the context of SMEs' digital supply chain management. *Production Planning & Control, 28*, 1378–1388.

Sodhi, M. S., Tang, C. S., & Willenson, E. T. (2021). Research opportunities in preparing supply chains of essential goods for future pandemics. *International Journal of Production Research*. https://doi.org/10.1080/00207543.2021.1884310

Velayutham, A., Rahman, A. R., Narayan, A., & Wang, M. (2021). Pandemic turned into pandemonium: The effect on supply chains and the role of accounting information. *Accounting, Auditing & Accountability Journal, 34*, 1404–1405.

Wang, M., Wang, B., Song, B., & Bayne, K. (2020). *Understanding effects of COVID-19 in New Zealand businesses: An early-stage study*. Scion: Business Outlook June: 1–8.

Yu, K. D. S., & Aviso, K. B. (2020). Modelling the economic impact and ripple effects of disease outbreaks. *Process Integration and Optimization for Sustainability, 4*, 183–186.

Zidi, S., Hamani, N., & Kermad, L. (2021). New metrics for measuring supply chain reconfigurability. *Journal of Intelligent Manufacturing*. https://doi.org/10.1007/s10845-021-01798-9

Chapter 8
A Review of Supply Chain Management Practices: The Case of Botswana State-Owned Enterprises (SOE)

Oxford York and Dennis Sebata

Abstract COVID-19 has created many economic disruptions in the way a supply chain (SCs) is being managed. The rapid spread of the contagion has impacted tremendously the meat processing, food service and poultry SCs. Botswana being an underdeveloped country adopted supply chain management best practices to stay afloat during the pandemic. The lessons learned from Botswana can be used in developing countries to highlight the needed essential SCs elements during a pandemic. Botswana strong competitive advantages during crisis appears to be: service, operations, inbound and outbound logistics.

Keywords Food chain · Medical equipment · Processed meat · Health care supply · Essential worker shortage · Labor costs · Production demand · Pandemic leadership · Logistic innovation · Decision-making

8.1 Introduction

Worldwide attention was given as the breaking news about the coronavirus disease 2019 (COVID-19) viral outbreak unravelled in Wuhan, China. The live video footage and storylines being shown were unsettling, provoked fear and panic as the occurrence unfolded. Many researchers identified through environmental samples that the starting point of the new strain COVID -19 outbreak was in Huanan South China Seafood market (Gralinski & Menachery, 2020; Wang et al., 2020). According to the research of Ali et al. (2020), earlier cases revealed that only 22% of patients had direct exposure to the marketplace, 32% were in contact with the suspected cases and 51% had no contact with either of the source. In addition, there has not been any direct or identifiable association to any animals (Granlinski & Menachery, 2020). COVID-19 appears to have had a negative impact and influence upon one

O. York (✉) · D. Sebata
Monarch Business School Switzerland, Zug, Switzerland
e-mail: dr.york@monarch-university.ch; dr.york@umonarch-email.ch

D. Sebata
e-mail: dennis.sebata@umonarch-email.ch

H. Qudrat-Ullah (ed.), *Understanding the Dynamics of New Normal for Supply Chains*, Understanding Complex Systems,
https://doi.org/10.1007/978-3-031-07333-5_8

the most important and invaluable factors in the supply chain, which is the human factor. Being in a high state of urgency and ambiguity with no clear end in sight, one might question what are the short and long-term consequences of inappropriate and inaction of not finding rapid, optimal and adaptive solutions to sustain effective supply chain management. The former statement should elicit emotions of concern and worry on how to maintain resiliency, safety and flexibility within supply chains in order to adapt to such an impactful worldwide change.

There are no countries that was left unharmed by COVID -19. There are many questions that remain about the emergence of the COVID-19 but having a clearer comprehension of the evolutionary path may help to adopt better strategies and resolutions in suppressing the progress. Proper decision-making is essential in avoiding disruptions to supply chains, and to lessen any health risks that requires informed and diligent measures to be employed in order to constrain the propagation of COVID -19 for maintaining safe productivity and distribution within supply chain management strategies. This leads one to question how does one manage the supply chain for uncertainty and ambiguity or newness within a situation that appears to have transitioned from a pandemic into a possible endemic.

8.2 Supply Chain Management (SCM)

Supply chains seem to be constantly evolving, and appears to be a central discipline to comprehend management strategies. Supply chains are not impervious to disruptions from internal or external threats that may occur to the systems that are in place. In other words, it is not new for a supply chains to be forced to adapt to crisis situations. COVID-19 appears to present both short and long-term unforeseen effects on the way supply chains are managed. Supply chain management (SCM) can be explained as the necessary actions taken to coordinate production, location, inventory and transportation among the partakers within the supply chain to achieve effective and efficient responses for the particular market being served (Quynh & Huy, 2018). A supply chain encompasses the planning and management of all activities involved in sourcing and procurement, conversion and all logistics management activities including intermediary suppliers, third parties and customers (Blanchard, 2021). The framework of supply chain management that will be used for later discussion is based upon the five processes of Porter's competitive advantage theories:

1. *Inbound logistics.* These are the activities associated with receiving, storing, and disseminating inputs to the product (material handling, warehousing, inventory control, transportation scheduling, and returns to suppliers);
2. *Operations.* This refers to the activities associated with transforming inputs into the final product form (machining, packaging, assembly, equipment maintenance, testing, printing, and facility operations);

3. *Outbound logistics*. These are the activities associated with collecting, storing, and physically distributing the product to buyers (finished goods warehousing, material handling, freight delivery, order processing, and scheduling);
4. *Sales and marketing*. Within a supply chain context, these are the activities that induce buyers to purchase a product and enable them to buy it (advertising, promotions, sales force, quoting, channel selection, channel relations, and pricing);
5. *Service*. This refers to the activities associated with providing service to enhance or maintain the value of the product (installation, repair, training, parts supply, and product adjustment) (as cited in Blanchard, 2021, pp. 7–8).

The SCM core elements are expanded upon to fit with advancements in technology and customer demographics and new innovations in technological advancements and customer demographics, attributes and needs change (Min et al., 2019). The underlying goal of supply chain management is to be able to:

1. Identify the supply chain and its constituents;
2. Identify bottlenecks that are slowing down the movement of information of goods and services;
3. Having the right processes in place to get the right products delivered at the right time; and
4. Empowering the right people so they can accomplish all of the previously mentioned points (Blanchard, 2021).

Now that a framework for supply chain management has been discussed, the focus will switch towards exploring how the COVID-19 pandemic presents risks for all supply chains. The COVID-19 pandemic appears to have created side effects on the mental health, physical well-being, and the way of earning a living for individuals working within supply chains. Furthermore, the pandemic had dramatic effects on workers and their families, and businesses worldwide, especially small and medium sized enterprises (United Nations, 2020). Many workers had lost their way of earning a living due to job cuts from business closures or bankruptcy or because of the health risks and lack of personal protection equipment to provide safe practices (United Nations, 2020). Approximately 94% of worker's around the globe where residing in countries with some form of workplace closure protocols in effect (ILO, 2020). Despite of more countries easing restrictions, 20% of the workers around the globe resided in countries that required workspace closures except for essential workers, whereas, an additional 69% of workers lived in countries that required closure for some sectors or categories of work (ILO, 2020). The devastating impact of COVID-19 varied considerably between countries and groups of people based on the pre-existing government interventions and inequalities (United Nations, 2020).

Current updates on the labour market indicate a slow progressive return to work within distribution, manufacturing and production within the supply chain. Part of this progression has occurred due to vaccination as an imperative factor for labour market recovery (ILO, 2021). However, there seems to remain a major discrepancy between high-income and low-income countries in terms of accessibility and planned actions for receiving vaccinations. Being that COVID-19 is a novel disease there are no vaccines or highly effective treatments available, therefore, the virus continues to spread (Rejeb et al., 2020). There appears to be national interests from government officials in vaccinating most of the population as a solution for containing the disease, which is now considered as the reality of "the new normal" (Kersan-Skabic, 2021). Countries imposed severe lockdowns measures to lessen the spread of the virus and to prevent a collapse of the health care system from occurring, but none of these actions have proven to be highly impactful in recovering from the situation (Rejeb et al., 2020). The research indicates the possibility of COVID-19 becoming an endemic and seasonal (Calina et al., 2020). Further review is required to comprehend how different supply chains are managing with the above arguments.

According to the International Labour Organization, 59.8% of high-income and 1.6% of low-income countries have received vaccinations to allow for a return to the workplace (ILO, 2021). Interestingly enough, having less rigid work restrictions are associated with higher vaccination rates (ILO, 2021). How does the former statement make sense? It can appear that specific areas or sectors are being targeted for closure. Additionally, the decision-making within the policies and protocols in the labour market do not appear to make clear sense. It has been difficult for the global labour market to recover when a pronounced discrepancy exists in high and upper-middle income countries who have recovered but the lower-middle and low-income countries continue to suffer at a large loss (ILO, 2021). The former statement sheds light on the notion that lower paid workers and lower productivity businesses were inexplicably damaged by the pandemic. The labour market slow growth in productivity indicates a negative growth in low-and lower-middle income countries (ILO, 2021). The average worker in high-income countries produced 18 times more output per hour than the average worker in a low-income country (ILO, 2021). The COVID-19 pandemic has created the largest productivity gap seen since 2005 (ILO, 2021). The result of the former statement is an increased productivity gap between advanced economies and developing countries. Moreover, the pandemic may have shifted and stunned financially many low-income workers in developing countries creating labor shortages. One may question about what alternative strategies could be utilized by Small and Medium Size enterprises (SMEs) within low-income countries to help workers to still earn a living and to recover from this situation.

The COVID-19 complicated international production due to multiple barriers: border closing, shortage of health risks to staff, reduction in demand, income, job uncertainties and many others (Kersan-Skabic, 2021). The COVID-19 pandemic altered cross-border business and trade flows making this an important side effect to be examined on how the main issues and challenges are being addressed in international trade and business suppliers (Kersan-Skabic, 2021). The former statement implies major deficiencies and shortages within sectors or sections of the supply chains and the need for alternative means. International production depends highly on exportation and importation of immediate and final products plus trade policy rules governing the trade (Hayakawa & Mukunoki, 2021). For example, the negative effects of COVID-19 has impacted international trade on non-essential products and had a positive effect on providing medical products (Hayakawa & Mukunoki,

2021). The COVID-19 outbreak has created change within the structure of the trade network. However, it is important to note that not all countries were impacted, such as China, who was able to maintain centre position in the trade network (Kersan-Skabic, 2021). Another question posed may be based on the type of supply chain management (SCM) strategies implemented or maintained to enhance or recover profitability and remain competitive. The subsequent section will discuss the impact of COVID-19 within the African continent.

8.3 Africa and COVID-19 Pandemic

During the onset of the COVID-19 pandemic Africa appeared to be safe, however, at a slow pace confirmed cases started to appear in the Northern, Southern and Western parts of the continent (Obande et al., 2021). The variation in transmission rates appeared to be influenced by socioeconomic status, nutrition, age, race, presence or absence of comorbidities (Prevent Epidemics, 2021). The African region had the third largest amount of cumulative deaths worldwide due to COVID-19 and at the same time, had the lowest number of tests per 100,000 persons since the start of the pandemic (Obande et al., 2021). The largest number of cases occurred in southern African region with 2,320,199 confirmed cases and 68,160 deaths (Obande et al, 2021). The landlocked Southern Africa country of Botswana conducted the highest amount of testing on the African continent by any country since the commencement of the pandemic (Obande et al., 2021).

Strict measures continue to be taken as the number of fatalities increased in some regions of Africa, this indicating a possible continued upward turn in the rate of infection within a continent where the healthcare system is not strong enough (Prevent Epidemics, 2021). Being that Southern Africa was hit the hardest during the pandemic, what were there recovery strategies? if any? The response to these questions will lead to learning the types of innovative practices utilized in Africa to help lessen the disease, especially in underdeveloped and more impoverished areas, such as Botswana, to allow for continuity of services and productivity through supply chain management best practices. There appears to be little to no research on how Botswana is faring in the pandemic and what adoption strategies are being utilized to maintain resiliency and avoid disruption in their supply chain. It appears that most of the extensive studies done have been on more developed countries but studying how underdeveloped countries have adapted to change can bring forth some new and innovative low-cost strategies and insights on coping with this pandemic with very little resources.

According to Setino and Amba (2016), South African government's supply chain management is not adequately implemented in state -owned enterprises (SOEs). There are apparent fragilities in the SOE's supply chain management enablers, the strategy, policy implementation and poor enforcement of government supply chain management rules and regulations (Sentino & Amba, 2016). Government officials should be more strategic around Supply chain management practices to improve delivery during the COVID-19 pandemic. The upper level management of SOEs do not appear to see the importance of giving SCM any attention and therefore, there is lack of support creating difficulty for supply chain management practitioners to execute their day to day functions (Sentino & Amba, 2016). This is indicative of misalignment and created blockages between organizational strategies and supply chain, which may pose greater challenges during the COVID-19 pandemic. There is a need for better services in Southern Africa to alleviate service delivery backlog and lessen any possible corruption from having a more solid structure (Sentino & Amba, 2016).

The current discussion has been understudied within the literature and any answers or hints should lead to a transfer of knowledge on what policies and procedures are critical within ambiguous and uncertain situations to sustain supply chain management within underdeveloped parts of the world. Furthermore, answers to the former questions can inform decision-making strategies for more developed countries. Botswana state owned enterprise (SOEs) will be used as a case study for implementation of essential supply chain management best practices during COVID-19. The reason for selecting Botswana is because it is an interlocked country within South Africa, meaning that it is surrounded and bordered by other countries presenting many challenges and risks to disruption within supply chain practices. Being that this appears to be an understudied subject area, an integrative and comparative literature review will be used to retrieve answers to the following main question for discussion:

> What are the essential supply chain management best practices being used in Botswana's state-owned enterprises during the pandemic to prevent a collapse of the existing supply chain?

The subsequent sections will explore the Africa continent supply chain disruptions due to COVID-19 and will also review standards of SCM practices in developing countries.

8.3.1 COVID-19 Impact in Africa

The number of fatalities in Africa due to COVID-19 appears to be low in comparison to more develop countries (OECD, 2020). However, COVID–19 has created economic distress in the following three areas: lower investment and traded from china in the immediate term; lockdowns created a demand decline in European and OECD countries; domestic and intra-African trade was impacted due to a continental shock in supply (OECD, 2020). It is important to note that underdeveloped African countries do not have viable alternatives to China as a buyer (OECD, 2020). Other sources of foreign direct investment into Africa seems to comes from the United States and France. The estimated earnings in Africa within the first months of the crisis had declined by approximately 80% in Africa (United Nations, 2020). The global COVID-19 disruptions will impact supply chains, and will lead to a decrease in the availability of final and intermediate goods imported to Africa (United Nations, 2020). The experience of relative poverty will seemingly increase in Africa as a long-term consequence of the pandemic.

The experienced low trade or no trade occurred when countries began to close borders for trade. Informal economy workers appear to have been hit the hardest in the pandemic. The informal economy can be defined as, "a process of income generation characterized by one central feature: it is unregulated by the institutions of society, in a legal and social environment in which similar activities are regulated" (Bromley & Wilson, 2018, p. 4). A large negative impact on informal workers incomes increased as their work exposed them to increased health and safety risks through a lack of proper personal protection equipment (PPE) and having to maintain strong interaction with co-workers (United Nations, 2020). This discussed vulnerability was experienced by informal economy workers in particular due to lockdown measures (Bromley & Wilson, 2018). In addition, the concentration of women in service provision sectors were more at risk than men due being able to do certain tasks that were better suited for a smaller female hand, and women appeared to be hit the hardest due to the lockdown measures (United Nations, 2020). This leads to question the adaptive measures used to maintain a resilient and flexible supply chain during these challenges. Furthermore, the uncertainty that came with the closure of businesses appear to have meant that industries were operating with limited resources or came to an end. The subsequent section will discuss Botswana and the COVID situation.

8.3.2 Botswana

In 1966, Botswana's independence marked tremendous socio-economic and political transformation that resulted in an accurate model of liberal democracy on the African continent (Ruele, 2011). The government of Botswana appeared to have diversified its economy since independence. Since independence, the country experienced rapid economic growth when customer satisfaction became a national priority (Ruele, 2011). Botswana has had the fastest growth per capita income globally, with an average growth of 9% per year (Harvey & Lewis, 1991). Botswana had gained a strong reputation for its economic management and adherence to democratic principles based on policy development to improve citizens' quality (Hope, 1995). The government has maintained a fiscal policy, which earned it the highest sovereign credit rating in Africa based on its impressive economic record (Ruele, 2011). However, it builds on the widely-used revenue generated from diamond mining for economic development through prudent fiscal policies and conservative foreign policy (Harvey & Lewis, 1991). More specifically, this was achieved through supply chain responsiveness, flexibility, leanness, agility, and efficiency synthesising in the public sector (Hallavo, 2015).

Botswana continues to have a significant problem post-independence due to poverty (Ruele, 2011). There are seemingly other factors that made it difficult to fight poverty, such as, landlessness, gender and ethic disparity, human immunodeficiency virus (HIV) and now a COVID-19. Also, Botswana is experiencing shortages in vaccinations for essential workers (United Nations, 2020). Despites all of these mentioned challenges, Botswana seems to have achieved remarkable growth in the economy, socio-political stability, and education with its current political system as one of the leading democratic regimes in Africa based on the equity in the distribution of resources and services (OECD, 2020). Botswana is an Africa role model of the free market economy, with a series of steps demonstrating good governance, liberalising and deregulating its economy, and removing tariffs and abandoning restrictions on exchange control (Marobela, 2008). Botswana economy appears to be enduring the COVID-19 pandemic as it relies entirely on government expenditure by either employment, procurement of good and services, income and tourisms. The research supports the notion that there exists a strong relationship between Botswana's organizational performance and supply chain management best practises (SCMBPs) (Kumar & Kushwaha, 2018). The formerly mentioned statement indicates that SCMBPs encourages growth, development, and economic integration through eliminating all tariffs and non-tariff barriers amongst its members (Cheong et al., 2018).

The Botswana public sector is comprised of three sub sectors:

1. Public Service (i.e. all Ministries and Independent departments);
2. Local Authorities (i.e. local government service catered by ministry) and
3. Parastatal sub-sector (i.e. quasi-governmental organizations and government agencies) (Hope, 1995).

The public sector's increased performance was attributed to competent staff members with integrity to fulfill their duties honestly and effectively (Quynh & Huy, 2018). Supply chain management success depends on each member being a customer and supplier for the strategy implementation (Kumar & Kushwaha, 2018). The concept of supply chain follows a logic to match with operational improvement (Hallavo, 2015). The subsequent section will explore the principle concepts behind supply chains in developing countries.

8.4 Supple Chain Management and COVID-19

Supply chains have been disrupted before by natural catastrophes that created worldwide distresses on supplies and distribution to meet demands. The major difference appears to be that the COVID-19 continues to spread with no apparent scientific evidence of a vaccine that works to slow down the contagion (Ali et al., 2020; Guo et al., 2020; Wang et al., 2020). COVID -19 is a reminder of the fragility of supply chain management systems (SCMs) and the imposed health risks on employment and income. Agri-food supply chains agents and economic actors will feel this sudden and long-term impact. The following section will discuss how the pandemic created disruption in the food supply chain disruption and effects of consumer choice.

8.4.1 Covid and Food Chain Supply

Global and local food systems were disrupted due to the COVID-19, and forced social distancing efforts (Niles et al., 2020). COVID-19 has disrupted food accessibility and has created food insecurity due to uncertainty on the safety of food products, individual and public health adverse consequences (Niles et al., 2020). Due to the numerous adverse health outcomes from COVID-19, food insecurity was developed (Niles et al., 2020). Food insecurity had occurred due to the lack of adequate access to food that meet dietary needs and the unknown health outcome risks to the public (Niles et al., 2020). At the start of the pandemic, the demand from restaurants fell drastically and created the need for many food products to be stored or further processed to avoid waste. The Canadian Dairy Commission borrowing limit increased to $200 million, and the Farm Credit Canada's lending capacity was increased by 5$ billion to help farmers deal with the cash flow and revenue lost (Larue, 2021). In other words, debt increased tremendously in these industries to adapt to the many sudden changes, thereby impacting the economy in terms of increasing pricing.

The resiliency of the food supply chain during the unfolding of COVID-19 requires careful attention. The rapid adjustment of food supply chains had to be implemented to deal with the demand-side disruptions, such as, the change patterns for in-food purchasing and panic buying (Hobbs, 2020). This also included planning for any supply-side shocks due to labour shortages and disruptions to transportation and supply networks (Hobbs, 2020). Panic buying and hoarding behaviors by consumers created a demand-side shock as governments worldwide increased the social distancing policies (Hobbs, 2020). The restrictions triggered fear and anticipation of a possible disruption to food distribution systems (Hobbs, 2020). Food supply chains disruptions were problematic as temporary closures for wholesale food supply chains, especially for food banks (Hobbs, 2021).

Agri-food trade costs increased and reduced in the competitiveness of cross border supply chains as trade movement restrictions occurred. The action of buying now for consuming later appears to creates a dynamic inventory problem (Cranfield, 2020). The shock to agricultural labor markets and production practices will trigger higher food prices in the long-term and volatility in price (Cranfield, 2020). All countries have particular groups that are vulnerable and affected by food insecurity created by the disruptions caused by COVID-19 (Cranfield, 2020). However, for example, Canada did not experience any restrictions on food and agricultural trade showing a resiliency on the continuity of food supply. The research in this area implies a need to maintain and enhance the resilience of supply chain through robust and reliable supply chain relationships (Hobbs, 2020). In Canada and United States of America (USA), food supply chains adapted well to the short-term halts in transportation and border closures. The resilience is emotive and a politically sensitive subject because it based on the interconnected supply chain of the broader concept of the food system (Hobbs, 2021). Some of these mentioned concepts, include the importance of prioritizing open borders for the flow of essential goods during a crisis; The second vulnerability involves labor, worker illness, labor shortages, self-isolation or movement restriction (Hobbs, 2020, 2021).

Food supply chain actors have proven to have an ability to respond at a remarkable speed in avoiding discontinuity of available food (Deconinck et al., 2021). The rapid response time is indicative of having supply chain flexibility, which includes accessible and predictable international trading environment to that allows for entering into new supply when existing sources experience compromise (Deconinck et al., 2021). The former statement seems to be true for developed countries but leaves to discussion the reaction time for undeveloped countries to sustain their food supply. Seasonal workers for planting and harvesting fruit and vegetables in many countries are at risk for delays in distribution (Deconinck et al., 2021). There is a risk of disruption for the transportation of seeds by air due to cessation of air transport by certain countries. These experienced limits in mobility reduced the local distribution because of the difficulty in transportation (Deconick et al., 2021). There are three noted blockages that require careful attention. According to Deconinck et al. (2021), policymakers should be made aware of: the availability of labour for harvesting fruits and vegetables, meat processing sector implications of the shutdown, and ongoing disruption of air freight within high-value perishable products.

As the COVID-19 disease continues and mutates, the research indicates this may pose new risks to global food supply chains may emerge. The largest threat to the food security comes from the devasting effects that covid has on livelihoods and jobs, especially with developing countries where safety nets are less developed and may create increase poverty and hunger (Deconick et al., 2021; Hobbs, 2021). Having a diversified source of supply seems to allows for rapid response during compromise by transport or logistics disruptions, having open and predictable markets are imperative for distribution of food along supply chains (Deconick et al., 2021; Hobbs, 2021). The following will discuss meat processing sector and how they cope during the pandemic.

8.4.2 Covid and Meat Processing

Meat processing plants had become the hotspots COVID-19 transmission. The virus appears to thrive in cold, and therefore, the refrigerated conditions in the plants were likely fueling the infection (Reid et al., 2021). The U.S. Department of Agriculture (USDA) initiated an audit on actions that may have contributed to the spread of COVID-19 in meat processing facilities (Reid et al., 2021). Some of the highest rates of COVID-19 infections came from meat processing plants (Fatka, 2021a). The impact of this spread early in the pandemic harmed the workforce that was compromised of immigrant's refuges, people of color (Fatka, 2021a). This stimulated questions about federal actions that may have led towards the virus in these facilities. The US implemented new and comprehensive policies to protect the meat sector. The new policies have lowered by five times the amount of cases, which is now considered lower than the general population cases, and down by 95% from the pandemics peak in the US (Fatka, 2021b). Frontline meat and poultry workers are being prioritized by government for vaccinations to make sure that Americans and the farmers economy have no disruptions in food supply (Fatka, 2021a). Hence, leading to increased food prices and the staggering cost of value of livestock for farmers (Fatka, 2021b). The following section will discuss about the impacts occurring disruption and deficiencies on the health care supply chains and the risks towards providers health.

8.4.3 Health Care Supply Chains and Covid

COVID-19 has created significant changes and unforeseen problems within the health care supply chains. There are five categories of products required in healthcare: blood, medical supplies, medical devices, PPE and pharmaceuticals (Mirchandani, 2020). Being that each of these categories have their own supply chain distribution any interruption in one of these areas will create havoc on the entire system. Pharmaceutical supply chains are global and many of the active ingredients are manufactured overseas, China provides two thirds of the active ingredients of generic drugs to the United States of America (US) (Mirchandani, 2020). Italy, Belgium and United Kingdom are some other countries involved in manufacturing active ingredients. Each of these areas where hit hard during to the pandemic, thereby, impacting supply distribution for active medical ingredients (Mirchandani, 2020). Many of the virus-related manufacturing issues and regulatory restrictions have created problems in the supply chains for pharmaceuticals. The US appears to have survived the disruption from having a large over stockpile of inventory, which a shortage could follow once the pipe line inventory is depleted (Mirchandani, 2020).

Health care providers treating COVID-19 patients should have been well protected but that did not appear to be the case. Health care providers were exposed high health risks carrying out their duties. Many countries limited manufacturing capacity would have taken months to fulfill the millions of required PPE, and quickly diminishing emergency backup supplies, such as, N95 respirators (Mirchandani, 2020). Medical devices tend to have highly regulated supply chains. Devices, such as, mechanical ventilators where limited and also brought forth uncertainty surrounding the effectiveness for usage against COVID-19 (Mirchandani, 2020). Furthermore, the lack of standardization of mechanical ventilator brands created more difficulties in getting replacement parts for the devise. Furthermore, medical supplies for testing materials, laboratories and intravenous kits and surgical center supplies. COVID-19 clearly disrupted and frustrated the supply chain response strategies. For example, there are only two companies responsible for supplying the swabs needed to collect testing samples (Mirchandani, 2020). The former statement seemed to be the catalyst that forced newer tests and alternate protocols to be made. Finally, COVID-19 negatively impacted the blood supply chain by making it problematic for donating blood. The shortage on blood supplies and social distancing made it difficult to collect blood (Mirchandani, 2020). Also, the overall fear of acquiring the virus prevented blood donors from giving volunteer time (Mirchandani, 2020). COVID -19 appears to be driving the health care system supply chains to re-evaluate and become stronger positioned to handle reoccurrences for future pandemics. The subsequent section will discuss meat processing industry and the work conditions of these essential workers.

8.4.4 Covid and Essential Workers

The research opens up the discussion on what type of critical resources were available to the workers in order to offer a safe work environment. An ethical dilemma appears to be raised not only on the safety of the workers but on how and where government/businesses were allocating funds to protect workers, provide healthy environment conditions and customer safety of food and meat products within the supply chain. Essential migrant and immigrant workers have bared the weight of the COVID-19 (Reid et al., 2021). There appears to have been inequitable work conditions. The term essential workers have been independent in providing safety resources and protection, they tend to be lower paid and unentitled to paid sick leave, thereby, having to work despite of being infected by COVID-19 (Reid et al., 2021). A more positive outcome of the COVID-19 pandemic can occur if living and working conditions of migrant workers improve (Reid et al., 2021).

There appears to be similar dilemmas occurring in other countries outside of North America in the meat processing industry. In the worldwide production of poultry, beef and pork, transnational networks of corporations have attracted public investment at source or subsides (De Campos Silva, 2020). There is a history of worker abuse in the poultry sector that has led to a range of occupational health issues, musculoskeletal diseases and mental health issues (De Campos Silva, 2020). Women poultry workers experience more occupational health hazards, and are deemed as better suited for operations in this industry that requires a particular manual dexterity not seen in men (De Campos Silva, 2020). Packers and slaughterers tend to be in degrading work conditions in this industry, despite of, the invest of biotechnology and automation (De Campos Silva, 2020). In the halal sector, there are the chicken bleeders with a focus on the Muslim markets and must be carried out by Muslim men (De Campos Silva, 2020).

The Brazilian poultry began hiring asylum seekers within the industry and this concept is not foreign to North America. Poultry appears to be one of the most affordable proteins in the world and when there is a disruption it can have an impact on human nutrition. In Brazil, COVID-19 is attracting more attention to the existing poor work conditions in the poultry industry, and it has also become a hot spot for infection (De Campos Silva, 2020). Workers are not provided with proper personal protection equipment (PPE) and have to work in close proximity (De Campos Silva, 2020). Two major US conglomerates in Brazil lobbied the government to remain open during the pandemic causing three deaths and spreading the virus to small towns (De Campos Silva, 2020). Brazil poultry processing became virus spread-breeders. It has reached to the point where Brazil may be experiencing the construction of a proto-pandemic ecology as it promotes the very practices that create pandemics and endemics to commence (De Campos Silva, 2020). The global emergence of new pathogens will occur as the poultry industry depends on genetic monoculture of chickens (Hafez & Attia, 2020). It is important to note that chickens are not susceptible to intranasal infection by COVID -19 virus (Hafez & Attia, 2020). However, COVID-19 will have an impact on poultry farming, transportation and consumption.

International migrants are among the socially vulnerable groups in terms of transmission of COVID-19 (Diaz et al., 2021). Migrant workers are overrepresented in COVID-19 laboratory tests, hospitalizations and deaths (Diaz et al., 2021). In order to reduce inequalities in this disease burden, there needs to be put into effect counterfactual policies to comprehend the underlying mechanisms behind this issue (Diaz et al., 2021). There is an apparent need for more research in this subject as it poses many unanswered questions and hints to an existing gap in the literature, meaning critical research that is not being addressed or focused upon. This is an indication for better decision-making for workers health and safety to offer better work conditions and health precautions, as they are an essential component of the supply chain. There is no clear reasons or explanations given in the literature but leaves one to speculate on why are there more health risks for migrant workers as opposed to the host population.

8.4.5 Unemployment During the Pandemic

In 2020, Canada marked a one-year increase within the unemployment rate. Low-wage workers and restaurant workers were hit the hardest during the pandemic. The essential service of food distribution and agri-food supply chains proved to be resilient against the public health measures and the pandemic (Larue, 2021). However, the labour market was disrupted by the enormity of COVID -19 as previous discussed in other countries. The forced public health restrictions severely diminished the demand for workers (Larue, 2021). Furthermore, the uncertainty of the pandemic outcomes created questions about how many businesses would have to close down. The governments were forced to make rapid decisions to slow down the rate of infection. Between February and May of 2020, the unemployment rate in Canada was 13.7 and 14.7% in the US (Larue, 2021). In accordance to North American standards, Some European countries experience a slight increase in unemployment and had less stringent lockdown measures and steadily continue to decrease in rate during the months of May, June and July 2020 (Larue, 2021). The lockdown measures cannot seemingly be the sole reasoning for the increase in unemployment but perhaps through the decision-making conflicts and needed rapidity of implementing policies and plans to help eradicate the virus and keeping everyone safe. Policy makers had to experienced tremendous pressure in decision- making within the ambiguous and unpredictable situation.

As stated by Larue (2021), “It is easy to criticize the policy response in Canada and elsewhere, but policymakers were under tremendous pressure and had very little information to rapidly adjust current policies and programs and to design and to implement new ones" (p. 272). Governments around the world experienced shortages in medical supplies and testing equipment for COVID-19 and had to make on the spot decisions being that there was limited scientific information about the virus (Larue, 2021). European policy makers seemed to have placed more decision-making emphasis on their regulations for protecting employment, which explains the drastic differences in the unemployment rate to North American countries (Blanchard & Portugal, 2001; Larue, 2021). The pandemic on a positive note seemed to have created more jobs for those who are able to work from home, and accelerated the digitalization of the economy. Conversely, many jobs were lost in this process of digitalization or had become obsolete (Deady et al., 2020).

Businesses, governments and nations have been forced to focus on the economic, financial and social implications created by the COVID-19 pandemic (Bhattachary et al., 2021). Organizations had to quickly adapt to the crisis by implementing new work from home strategies, communication new work arrangement, such as, working from home as new practice for many organizations. Institutes and industries have observed crises in the past and have adapted as a preventative measure in case similar occurrences happened in the future (Bhattachary et al., 2021). The COVID-19 pandemic created highly volatility markets and corporate failures due to financial instability, and scandals due to lack of leadership to effectively manage the crisis (Bhattachary et al., 2021).

In Canada and many other countries, there was a visible shortage in essential items (e.g. N95 masks, gloves and sterilization products) and this led to an increased demand spike that triggered panic buying and hoarding behaviors by consumers (Clapp & Moseley, 2020; Hobbs, 2020). There has been continuous world debates on how this crisis has been managed, the appropriateness of policy responses and weaknesses within the current system (Clapp & Moseley, 2020). Furthermore, there has been a large impact on labor shortages include health and illness of the workers, mental health in terms of self-isolation and movement distribution restrictions (Hobbs, 2020). It will be necessary to comprehend, knowledge share and attain on how decision-making will proceed Post COVID-19. The following section will discuss supply chain management practices during COVID-19.

8.5 Supply Chain Management Practices

The COVID-19 pandemic is a reminder of the fragility and sensitive nature of supply chain networks around the world and places a focus on the risks and reality of the flow of consequences that can occur within multiple system failure (Rejeb et al., 2020). The management of supply chain disruption poses many challenges but how it is being managed indicates robustness of the supply management system, which includes a contingency plan for mitigating risks and monitoring the system during a period of disruption (Vieira, 2020). The covid pandemic appears to have impacted supply chain management processes from the economies of scale, meaning the efficiencies in the vendor relationships, inventory control, logistics and production (Hobbs, 2021; Rejeb et al., 2020). Furthermore, it appears that COVID-19 may have impacted the economies of scope, meaning the increasing return from Supply chain management needed to facilitate and move product at a greater pace (Esper, 2021). In other words, the conversion of materials and components to be converted into finished products and the logistics to get those products to market (Esper, 2021). The failed ability of supply chains to get products into the market during the pandemic created worldwide media attention.

Initially, the attention of the coronavirus appeared to be an issue impacting and affecting China. The reality of the situation was that 95% of Fortune 1000 companies were already impacted due to having the global supply chain operations in China and naturally, this created interruptions for inventory and direct product movement (Esper, 2021). It appears that the pandemic has initiated global supply chain management (SCM) risk conservations that could not have been brought to forefront without this occurrence (Esper, 2021). The SCM risk research appears to focus mainly on the operational risks that impact and posed threat on inventory investment and the cost of the supply chain (Sodhi et al., 2012). All supply chains are vulnerable to risks and disruptions, COVID -19 appears to place an uncertainty about the readiness for another future event that resembles this one. Moreover, the pandemic appears to not have a foreseen end point and is denoting a new norm as organizations, businesses and SCM policies and procedures continue to move forward. These mentioned implications have also caused companies, such as Amazon, to experience publicized criticism towards the work conditions within their processing plants and distribution centers for their supply chain (Esper, 2021). It appears that supply chain workers understand that they are not immune to the stress of COVID-19 and is hoping a resolution can be found to improve work conditions and stop or lower the spread of the virus (Rejeb et al., 2020). This former statement leads to questions about what will be adapted to protect the health and safety of frontline workers Post COVID-19.

The discussion thus far has been about well-developed countries and how they have managed and, in some cases, have begun or entered into the recovery phase and seen adaptiveness in the SCM processes from the COVID-19 initial impacts. One lingering question remaining is how did undeveloped countries fair in the pandemic and the strategies used with little too few resources. In the following sections, Botswana will be focused upon to explore decisions made during COVID -19.

8.6 Implementing SCM Strategy in Botswana During COVID

There are many challenges faced by Botswana for implementation of effective supply chain strategies. Hence, placing Botswana in a disadvantage when disruptions occur it seems. Firstly, the unavailability of technological expertise for equipment, human capital, and technology puts developing countries like Botswana much behind more developed countries (Ben-Daya et al., 2019). Secondly, non-existent supplier relationships and over-saturation in the supplier market may have led to poor relations between the supply chain and their respective suppliers. The former statement may lead to problems down the supply chain line for the entire process (Ozkan-Ozen et al., 2020). Thirdly, the lack of basic infrastructure is an enormous hurdle to cross in underdeveloped or developing economies. One can imagine that these challenges are even more difficult when measured economically. For example, the supply chain costs in Botswana are much higher than that in the US. These increased costs translate to an imperfect system providing lower value to all concerned parties (Castillo et al., 2018). The following section will take an in-depth look at what was done in Botswana to learn essential lessons.

8.6.1 Relationship Management in Botswana State Owned Enterprises

Botswana has worked on its macro-economic policies to attract state-owned companies and agencies (Owusu & Ismail Samatar, 1997). Botswana has also placed focused on developing its industrial sector, including manufacturing and service (Barclay, 2002). The government of Botswana also brought manufacturing and the service sector into competitive standards (Chiguvi, 2020). State-owned enterprises are pushed to deploy company resources to increase product and service experience quality for the customers to achieve customer satisfaction (Rapitsenyane, 2019). Some critical steps have been taken by the government of Botswana that has contributed to the growth of state-owned enterprises, such as, a change of policies, innovative management practices, and effective integration of information technology and customer relationship management (Chiguvi, 2020). The government of Botswana seems to puts a sincere effort to increase competitive measures in their indigenous industries to attract business development and strengthen the economy (Chiguvi, 2020; Sebata, 2021). To have effective customer relationship management, state-owned enterprises of Botswana followed advancements in technology using integrated software and innovative management practices (Sebata, 2021). They also worked to achieve effective communication between the channel partners, enhancing service standards for the customers (Sebata, 2021).

According to Sebata (2021), the misalignment of supply chain practices within the Botswana public sector during the COVID-19 pandemic was a result of a poor focus on ownership structures and commercial outcomes. During the COVID - 19, supply chain management best practices in Botswana switched focus to allow SOE's procurement policies to acquire materials within government partnerships only (Sebata, 2021). Certain postponement strategies adopted in Botswana's SOEs appears to be implemented to optimize the costs by partners. Furthermore, some of Botswana's postponement strategies are logistics done by various supply partners of parastatals and manufacturing (Sebata, 2021). The state-owned enterprises use postponement practices to minimize waste and meet customer demands. Moreover, Botswana has developed an informed supply chain management practices as the country was able to make an entrance into industries of developing countries. Botswana implements full postponement and majority manufacturing and full specification postponement strategies.

The local indigenous citizens and the community, get a greater understanding of the overall performance of the public sector undertaking. Public servants in Botswana contributes to the delivery of services to consumers and partnerships between the national affairs and public sector to enable the county to acquire a reputation for good governance and development engagement for efficient customer relationships. The government has made efforts in transitioning during the COVID-19 their marketing strategy and customer relationships strategies from mass appeal to customizable marketing to satisfy the customers, under clients need sand predict the future requirements of the market (Sebata, 2021). Botswana's strategic supplier relationship management practices are implemented on long-term relationships with suppliers or critical suppliers' involvement in continuous improvement programs, contributing to improving organization performance (Sebata, 2021).

In terms of Information technology, Botswana has not fully developed much advancement in this area. Therefore, it is depended on manual strategies to manage and align its supply chains with customer satisfaction and service quality (Sebata, 2021). Effective sharing occurs through sharing partners for business planning that is needed. This is indicative of trust and a sense of belongingness with the supply partners. This is how the organization and its trading partners keep each other informed about event or changes, then the organization 's performance showed improved during this period (Sebata, 2021).

8.7 Implications for Supply Chain Management

Going forward in a post-COVID world there are a few implications for businesses. The extensive indirect and direct damages from COVID-19 can seemingly be taken as a learning strategy in finding balance, resiliency and safety for continuity and improved productivity in preparation for the next crisis. This discussion highlights how rapid a contagion can spread through human contact or possibly supply distribution. More importantly, it highlights the importance of the human factor in maintaining a solid and flexible supply chain best practice during crisis. The short-term consequences of poor decision-making can be life altering as discussed in this chapter leading to death or long-term illness. COVID-19 has notably created an indirect disruption in the economy, labor shortages and current supply chains worldwide. The term indirect is being used as the issues appears to be decisions made due to lack of awareness and factual knowledge of the present situation. It appears to be based on the virus but the more problematic areas have more to do with the lack of informed decision-making processes being used world-wide. There appears to be a major ethical dilemma occurring behind the scenes of the existing chain management protocols and procedures leading to an increased spread of the contagion. One dilemma would be the nation's rapid response in purchasing vaccinations that have not proven to be effective in reducing the transmission, while exposing the workers to increased risk and possible sources of transmission from not having proper PPE and work conditions to lower health risks and maintain production on various supply chains. Better decision-making is needed.

The lessons learned from Botswana show that when adaptive measures are put in place, it helps ease crisis in order to maintain and implement best practices in supply chain management to stay afloat and remain competitive during the pandemic.

> What are the essential supply chain management best practices being used in Botswana's state-owned enterprises during the pandemic to prevent a collapse of the existing supply chain?

The answer consists of five elements:

1. Decision making that will be based on protecting the workers employment Health and Safety Education and Training and PPE;
2. Rapid Covid testing before and after exposure; Not only on entry border but within the community;
3. Support, trust, feeling of belonginess and a diverse partnership relationship, government help in order to have flexibility and adaptability of the chain; and
4. Constant communication with all parts of the supply chain.
5. Reshoring manufacturing and distribution of supply chains to improve the local survivability of each region with the understanding that it will have an impact on profitability and increased expenses.
6. Identify and evaluate barriers to reach a common conclusion.

The framework of Botswana's strong competitive advantages during crisis appears to be: service, operations, inbound and outbound logistics. The developed world can learn a great deal from this underdeveloped country. Decision-makers in Botswana supply chain best practices appear to have shown the importance of balancing trade-offs between survivability and excess demand. This chapter helps in learning about resiliency and flexibility within supply chains and logistics to cope with future disruptions.

References

Ali, S. A., Baloch, M., Ahmed, N., Ali, A. A., & Iqbal, A. (2020). The outbreak of coronavirus disease 2019 (COVID-19)—An emerging global health threat. *Journal of Infection and Public Health, 13*(4), 644–646.

Barclay, I. (2002). Organisational factors for success in new product development. *IEEE Proceedings—Science, Measurement and Technology, 149*(2), 105–112.

Ben-Daya, M., Hassini, E., & Bahroun, Z. (2019). Internet of things and supply chain management: A literature review. *International Journal of Production Research, 57*(15–16), 4719–4742.

Bhattacharay, S., Smark, C., & Mir, M. (2021). Covid-19: Social, financial and economic implications. *AABFJ, 15*(1), 1–4.

Blanchard, D. (2021). *Supply chain management best practices* (3rd ed.). Wiley.

Blanchard, O., & Portugal, P. (2001). What hides behind an unemployment rate: Comparing Portuguese and U.S. labor markets. *American Economic Review, 91*(1), 187–207.

Bromley, R., & Wilson, T. D. (2018). The urban informal economy revisited. *Latin American Perspectives, 45*(1) issue 128, 4–23.

Calina, D., Docea, A. O., Petrakis, D., Egorov, A. M., Ishmukhametov, A. A., Gabibov, A. G., Shtilman, M. L., Kostoff, R., Caralho, F., Vinceti, M., Spandidios, D. A., & Tsatsakis, A. (2020). Towards effective covid-19 vaccines: Updates perspectives and challenges (review). *International Journal of Molecular Medicine, 46*, 3–16.

Castillo, V. E., Mollenkopf, D. A., Bell, J. E., & Bozdogan, H. (2018). Supply chain integrity: A key to sustainable supply chain management. *Journal of Business Logistics, 39*(1), 38–56.

Cheong, J., Kwak, D. W., & Tang, K. K. (2018). The trade effects of tariffs and non-tariff changes of preferential trade agreements. *Economic Modelling, 70*, 370–382.

Clapp, J., & Moseley, W. G. (2020). The food crisis is different: Covid-19 and the fragility of the neoliberal food security order. *Journal of Peasant Studies, 47*(7), 1393–1417.

Cranfield, J. A. L. (2020). Framing consumer food demand responses in a viral pandemic. *Canadian Journal of Agricultural Economics., 68*, 151–156.

Deady, M., Tan, L., Kugenthiran, N., Collins, D., Christensen, H., & Harvey, S.B. (2020). Unemployment, suicide and covid-19: Using the evidence to plan for prevention. *The Medical Journal of Australia, 213*(4), 153–154.

De Campos Silva, A. R. (2020). Health risks for poultry workers in Brazil in the covid-19 pandemic. *Bulletin of Latin American Research, 39*(S1), 88–91.

Deconinck, K., Avery, E., & Jackson, L. A. (2021). Food supply chains and covid-19: Impacts and policy lessons. *EuroChoices, 19*(3), 34–39.

Diaz, E., Mamelund, S., Eid, J., Aasen, H. S., Kaarboe, O. M., Brokstad, R. J. C., Gloppen, S., Beyer, A., & Kumar, B. N. (2021). Learning from the covid-19 pandemic among migrants: An innovative, system-level, interdisciplinary approach is needed to improve public health. *Scandinavian Journal of Public Health, 49*, 804–808.

Esper, T. L. (2021). Supply chain management amid the coronavirus pandemic. *Journal of Public Policy & Marketing, 40*(1), 101–102.

Fatka, J. (2021a). USDA inspector to examine meat processing COVID cases. *Western Farm Press-Penton Business Media.* 1–3.

Fatka, J. (2021b). Higher food prices could result if meat plants required to increases space requirements between workers. *Western Farm Press- Penton Business Media.*

Gralinski, L. E., & Menachery, V. D. (2020). Return of the coronavirus: 2019 -nCov. *Viruses, 12*(135), 1–8.

Guo, Y., Cao, Q., Hong, Z., Tan, Y., Chen, S., Jin, H., Tan, K., Wang, D., & Yan, Y. (2020). The origin, transmission and clinical therapies on coronavirus disease 2019 (COVID-19) outbreak-an update on the status. *Military Medical Research, 7*(11), 1–10.

Hafez, H. M., & Attia, Y. A. (2020). Challenges to the poultry industry: Current perspectives and strategic future after the covid-19 outbreak. *Frontiers in Veterinary Science, 7*(516), 1–16.

Harvey, C., & Lewis, S. R. (1991). Policy choice and development performance in Botswana. *African Studies Review, 34*(3), 137–139.

Hallavo, V. (2015). Superior performance through supply chain fit: A synthesis. *Supply Chain Management: An International Journal, 20*(1), 71–82.

Hayakawa, K., & Mukunoki, H. (2021). The impact of covid-19 on international trade: Evidence from the first shock. *Journal of the Japanese and International Economies, 60*, 101–135.

Hobbs, J. E. (2020). Food supply chains during Covid-19 pandemic. *Canadian Journal of Agricultural Economics., 68*, 171–176.

Hobbs, J. E. (2021). Food supply chain resilience and the covid-19 pandemic: What have we learned? *Canadian Journal of Agricultural Economics., 69*, 189–196.
Hope, K. R. (1995). Managing the public sector in Botswana. *International Journal of Public Sector Management, 8*(6), 51–62.
ILO. (2020, May). *International Labour Organization (ILO) covid-19 monitor* (4th ed.). https://www.ilo.org/wcmsp5/groups/public/---dgreports/---dcomm/documents/briefingnote/wcms_745963.pdf
ILO. (2021). *ILO monitor: Covid-19 and the world of work* (8th ed.). International Labour Organization. https://www.ilo.org/wcmsp5/groups/public/---dgreports/---dcomm/documents/briefingnote/wcms_824092.pdf
Kersan-Skabic, I. (2021). The covid-19 pandemic and the internationalization of production: A review of the literature. *Development Policy Review, 00*, 1–15.
Kumar, A., & Kushwaha, G. (2018). Supply chain management practices and operational performance of fair price shops in India: An empirical study. *Scientific Journal of Logistics, 14*(1), 85–99.
Larue, B. (2021). Covid-19 and labor issues: An assessment. *Canadian Journal of Agricultural Economics, 69*(2), 269–279.
Marobela, M. (2008). New public management and the corporatisation of the public sector in peripheral capitalist countries. *International Journal of Social Economics, 35*(6), 423–434.
Min, S., Zacharia, Z. G., & Smith, C. D. (2019). Defining supply chain management: In the past, present and future. *Journal of Business Logistics, 40*(1), 44–55.
Mirchandani, P. (2020). Health care supply chains: Covid-19 challenges and pressing actions. *Annals of Internal Medicine,173*(4), 300–301.
Niles, M. T., Bertmann, F., Belarmino, E. H., Wentworth, T., Biehl, E., & Neff, R. (2020). The early food insecurity impacts of covid-19. *Nutrients, 12.*
Obande, G. A., Bagudo, A. I., Mohamad, S., Deris, Z. Z., Harun, A., Yean, C. Y., Aziah, I., & Singh, K. K. B. (2021). Current state of covid-19 pandemic in africa: Lessons for today and the future. *International Journal of Environmental Research and Public Health, 18*, 1–15.
OECD. (2020). *Covid -19 in Africa: Regional socio-economic implications and policy priorities.* https://read.oecd-ilibrary.org/view/?ref=132_132745-u5pt1rdb5x&title=COVID-19-in-Africa-Regional-socio-economic-implications-and-policy-priorities&_ga=2.195473004.1388414703.1636511120-1376897899.1636511120
Owusu, F., & Ismail, A. (1997). Industrial strategy and the African state: The Botswana experience. *Canadian Journal of African Studies, 31* (2), 268–299.
Ozkan-Ozen, Y., Kazancoglu, Y., & Kumar Mangla, S. (2020). Synchronized barriers for circular supply chains in industry 3.5/industry 4.0 transition for sustainable resource management. *Resources, Conservation and Recycling, 161*, 104986.
Prevent Epidemics. 2021. *Update on COVID-19 in Africa.* https://preventepidemics.org/covid19/science/insights/update-on-covid-19-in-africa/
Quynh, D. V. X., & Huy, N. H. (2018). Supply chain management practices, competitive advantages and firm performance: A case of small and medium enterprises (SMEs) in vietnam. *Journal of Modern Accounting and Auditing, 14*(3), 136–146.
Rapitsenyane, Y. (2019). A conceptual review of sustainable innovation: A driver for growing the manufacturing industry in Botswana. *Journal of Creativity and Business Innovation, 5*, 43–61.
Reid, A., Ronda-Perez, E., & Schenker, M. B. (2021). Migrants workers, essential work, and covid -19. *American Journal of Industrial Medicine, 64*(2), 73–77.
Rejeb, A., Rejeb, K., & Keogh, J. G. (2020). Covid-19 and the food chain? Impacts and future research trends. *Scientific Journal of Logistics, 16*(4), 475–485.
Ruele, M. (2011). Eradicating poverty and promoting dignity in Botswana through contextual theology of liberation: Challenges and prospects. *Journal of Social Development in Africa, 26*(1), 161–186.

Sebata, D. (2021). *Supply chain management best practices: An empirical research on the botswana state-owned enterprises.* (Unpublished doctoral dissertation). Monarch Business School, Switzerland.

Setino, R., & Amba, I. M. (2016). Supply chain management practices in state-owned enterprises environment. *Risk Governance & Control: Financial Markets & Institutions, 6*(4), 380–391.

Sodhi, M. S., Son, B. G., & Tang, C. S. (2012). Researchers perspectives on supply chain risk management. *Production and Operations Management, 21*(1), 1–13.

United Nations. (2020). *Policy brief: The world of work and covid-19.* 1–24. United Nations. https://www.un.org/sites/un2.un.org/files/the_world_of_work_and_covid-19.pdf

Vieira, A. J. (2020). Supply chain disruptions and challenges post covid 19 crises in Indian context. *Supply Chain Pulse, 11*(1), 22–23.

Wang, W., Tang, J., & Wei, F. (2020). Updated understanding of the outbreak of 2019 novel coronavirus (2019-nCov) in Wuhan, China. *Journal of Medical Virology, 92*, 441–447.

Chapter 9
Collaboration Model Between Buyer and Supplier: An Empirical Assessment of Indonesian Pharmaceutical Industry

Erlinda Nusron Yunus

Abstract This study examines the collaboration engaged by buyers and suppliers in the pharmaceutical industry in Indonesia. Preliminary studies show that Indonesia's pharmaceutical industry is growing steadily at over 10% and has a better growth rate than neighboring South Asian countries. Unfortunately, the availability of drugs at the retail level (hospitals, health centers, and pharmacies) is relatively low, while the much-needed medicines are often unavailable. The situation shows poor coordination between entities in the drug distribution network. This study will thoroughly examine drugs' scarcity from the point of view of coordination between entities in the supply chain. This fashion has become even more crucial with the COVID-19 pandemic currently hitting the world, including Indonesia, as one of the emerging economies in Southeast Asia. Due to the pandemic, the availability of medicines and medical equipment is essential for the community. Collaboration between companies in the pharmaceutical industry is one of the keys to its smooth running. This study obtains 52 company data about supply chain relationships, architecture, collaboration, and performance. Using path analysis, the study shows that the supplier–buyer business relationship improves supply chain architecture, further increasing supply chain collaboration. As the level of partnership improves, the firms improve their performance. These findings are significant in the current situation, where pharmaceutical companies, distributors-retailers, and local governments need to work together to ensure the supply of medicines for the community.

Keywords Supply chain collaboration · Suppliers · Buyers · Pharmaceutical industry · COVID-19 pandemic · Health expenditure · Supply chain management · Business relationship · Supply chain architecture · Performance · Path analysis · Indonesia · Medical equipment · Medicines · Public health · Supply chain visibility · Agility · Flexibility · Pharmaceutical manufacturers · Supply channels · Drug costs · Industrial Revolution 4.0 · Asia–Pacific region

---

E. N. Yunus (✉)
Operations Excellence Department, Sekolah Tinggi Manajemen PPM, Jakarta, Indonesia
e-mail: erl@ppm-manajemen.ac.id

H. Qudrat-Ullah (ed.), *Understanding the Dynamics of New Normal for Supply Chains*, Understanding Complex Systems,
https://doi.org/10.1007/978-3-031-07333-5_9

9.1 Introduction

The Government of Indonesia has enacted the National Social Security System (SJSN) since 2004 to realize the highest degree of public health following the mandate of the 1945 Constitution articles 28H and 34. SJSN is a form of social protection to ensure a decent basic life of Indonesian society, including assistance and health services, both preventive and curative. Through the Ministry of Health, the Government targets the entire population of Indonesia covered under the National Health Insurance Program (as part of SJSN) in 2019 (Kementerian Kesehatan, 2015).

This government plan provides a positive opportunity for the pharmaceutical industry to give medicines to Indonesian society. Indonesia's pharmaceutical sector, which consists of pharmaceutical manufacturing companies, pharmaceutical distributors or wholesalers (hereinafter termed as PBF), as well as retails (hospitals, public health centers, pharmacies, and drugstores), is an industry that grows significantly at 10% per annum from 2010 to 2015 (OECD, 2018). The Ministry of Industry reported that the turnover of Indonesian pharmaceutical manufacturing reached Rp 69.4 trillion in 2014 will expand to Rp 102.05 trillion by 2020 (Fig. 9.1).

Growth in the pharmaceutical sector is an implication of increased public health spending. However, as reported by the Indonesian Ministry of Industry, significant industry performance is not necessarily supported by an increased public health expenditure, down from USD 358 per capita in 2016 to USD 352 in 2017 (World Bank, 2021a). Even though the number increased in 2018 (USD 375 per capita), the indicator is still far below the health spending in the country's neighbors, e.g., Malaysia, Singapore, the Philippines, Vietnam, and Thailand. Nationally, the total public health expenditure in Indonesia, which is 2.87% of GDP in 2018, is still lagging behind neighboring countries: Malaysia (3.76%), the Philippines (4.40%), Thailand (3.79%), and Vietnam (5.92%) (World Bank, 2021a).

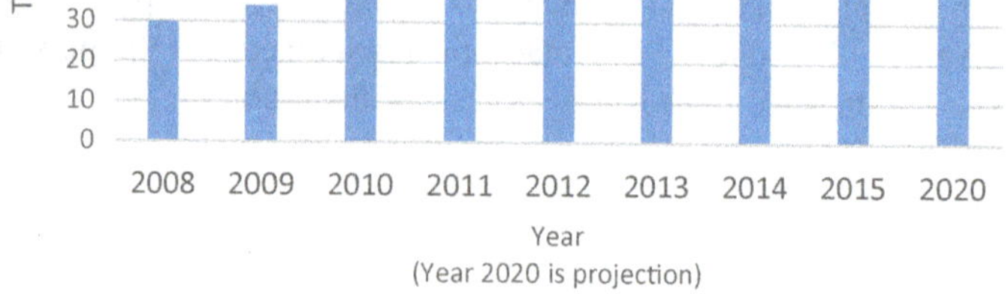

Fig. 9.1 Growth of the Indonesian pharmaceutical industry (*Source* Indonesian Ministry of Industry)

One of the causes of Indonesia's lagging compared to neighboring countries is the low accessibility of medicines in the community. People cannot shop for drugs because they are not available or not in good condition (Kementerian Kesehatan, 2015). The Ministry of Health reported that the availability of medicines and vaccines at the lowest medical-facility level only reached 75.5% in 2014. The Ministry's data also shows that inter-provincial inventory levels vary greatly, where there are provinces in 2012 with 80% drugs), while other regions reached more than 100% (overstock) (Kementerian Kesehatan, 2015).

Disparities that occur indicate not optimal supply chain governance (supply chain) in Indonesia. Unmanaged drug supply chain results in high (inefficient) shipping costs and uneven distribution (low drug accessibility). Therefore, a higher degree of collaboration is needed to ensure a smooth supply of medicines from manufacturing to retail, through the PBF and its branches nationwide.

So far, there has been minimal research on the collaboration or synergy of pharmaceutical governance in Indonesia. This study seeks to fill an existing gap by examining the effect of pharmaceutical company supply chain collaboration on company performance and overall supply chain performance. Furthermore, this study attempts to address some critical questions related to the collaboration or synergy of the pharmaceutical industry in Indonesia:

1. What is the level of collaboration or synergy of the pharmaceutical industry in Indonesia?
2. What is the relationship between business relationship, pharmaceutical architecture, supply chain collaboration, and supply chain performance in the pharmaceutical industry in Indonesia?

The findings will demonstrate the importance of supply chain collaboration in the pharmaceutical industry in Indonesia, that is, to support the accessibility of drugs to the community. The findings would benefit the industry and the government tremendously as they guide the pharmaceutical supply chain, which is very relevant to the current pandemic phase.

9.2 Supply Chain Management in the Pharmaceutical Industry

In Lambert and Cooper's (2000) study, The Global Supply Chain Forum (GSCF) defined supply chain management as an integrated business process, from suppliers to end consumers, by providing products, services, and information that add value to customers and all stakeholders. Previous research has shown the importance of entities in the supply chain to synergize to meet consumer needs as efficiently as possible (Ralston et al., 2017; Tarifa-Fernandez & De Burgos-Jiménez, 2017) and to mitigate the risks along the chain (Munir et al., 2020). Collaboration refers to the process of working together with others to produce something of common interest. Despite the inherent challenges of engaging in collaboration, companies strive for close cooperation to gain rewards, such as cost-effectiveness (Chen et al., 2017), lead time reduction and improved customer service (Al-Doori, 2019; Sheu et al., 2006), flexibility improvement (Chaudhuri et al., 2018; Danese et al., 2013) and increased profitability and competitiveness in the market (Flynn et al., 2010; Yunus & Tadisina, 2016). Recently, evidence also found that supply chain collaboration improves firms' sustainability performance (Chen et al., 2017).

The supply chain of the pharmaceutical industry is unique because it is tightly regulated. The relationship between suppliers and buyers in the pharmaceutical industry is bound by contracts and controlled by the relevant government agencies. Therefore, it is not easy for pharmaceutical manufacturers to determine who to collaborate with and to build long-term relationships that benefit all parties. Yunus and Kurniawan (2015) have examined the impact of a lack of coordination between a pharmaceutical company and its supply chain partners and proposed factors that influence this level of coordination. The study results show three primary triggers in successfully implementing buyer–supplier collaboration: trust, top management vision, and leadership. Leadership is the most relevant factor in explaining the unsuccessful partnership between two parties in the supply chain.

Furthermore, government plays an essential role in controlling a medical-related supply network. Yu et al. (2010) examined the performance and distortions in China's pharmaceutical industry at a macro level. The results of this research are in the form of a transition and economic reform in China, which impacts several problems, including ineffective supervision, price mark-up patterns, price scheduling distortions, and the absence of authorities to formulate drugs. The leading cause of market and government failures is the 'higher than cost' price demanded by all suppliers.

Although the supply chain in the pharmaceutical industry is distinct and challenging to manage, studies on supply chain management with this industry context are limited (Wang & Jie, 2019). In their study, Wang and Jie conceptualized that pharmaceutical companies need to hinder risk and uncertainty by improving their supply chain visibility, agility, and flexibility. Moreover, Moosivand et al. (2019) suggested that pharmaceutical manufacturers have better forecast accuracy and maintain an optimal inventory level.

9.2.1 Supply Chain for Medicines in Indonesia

Research by Mustamu (2007) described the supply chain activities of the pharmaceutical industry in Indonesia (as shown in Fig. 9.2). This industry has a long supply chain of suppliers, large pharmaceutical manufacturers, distributors (PBF), sub-distributor, and retailers. The pharmaceutical industry in Indonesia takes 120 days from upstream to downstream, that is, 60 days for production and 60 days for transportation. As a result of the length of the production chain, several losses can arise in reduced opportunities for products to be absorbed by consumers more quickly, and there is a risk of product damage due to limited expiry time (expired date).

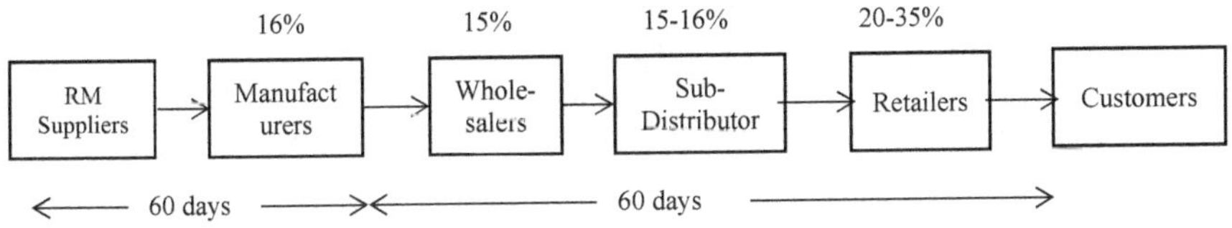

Fig. 9.2 Flow of pharmaceutical industry supply chain in Indonesia

In the context of the pharmaceutical industry, the process along the supply chain is dynamic. Therefore, controlling all supply channels is much more complex than in other manufacturing industries (Kiely, 2004). Therefore, the longer and more active the supply chain, the more critical forecasting, and demand planning activities will be. Mustamu (2007) provides solutions to improve the supply chain flow of the pharmaceutical industry by utilizing information technology in business processes along the supply chain. The application of information technology can shorten the delivery time up to 80 days. In addition to implementing the EDI process, e-commerce can also cut the "sub-distributor" chain, saving 15–16%. The latest solution is a benchmark from Japan, Malaysia, and Singapore, which implement self-dispensing (without going through pharmacies) by doctors, thus providing an opportunity for lower drug costs.

Given the importance of collaboration in a pharmaceutical supply chain, supplier and buyer relationships are the first to build long-term partnerships (Lee & Ha, 2020). Relationships that are mutually dependent, intensive, and based on solid trust will enable companies to carry out the initial stages of collaboration, namely sharing information and technology (Al-Doori, 2019; Ha et al., 2011). Sheu et al. (2006) proposed several essential factors that influence the level of collaboration, either directly or indirectly. These factors are business relationships (which can be measured by intensity, interdependence, and trust), long-term orientation, and supply chain architecture (represented by information sharing, inventory systems, information technology capabilities, and coordination structures). Based on these arguments, this study conjectures that,

H1 Supplier-Buyer Business Relationship has a positive relationship with the Supply Chain Architecture

H2 Supply Chain Architecture has a positive relationship with Supply Chain Collaboration

Furthermore, the previous discussion highlights the importance of increasing collaboration levels to have better supply chain performance (Al-Doori, 2019; Flynn et al., 2010; Yunus & Tadisina, 2016). Directly, supply chain architecture also plays a role in achieving higher performance (Saeed et al., 2019). Therefore, we posit that,

H3 Supply Chain Architecture has a positive relationship with Supply Chain Performance

H4 Supply Chain Collaboration has a positive relationship with Supply Chain Performance

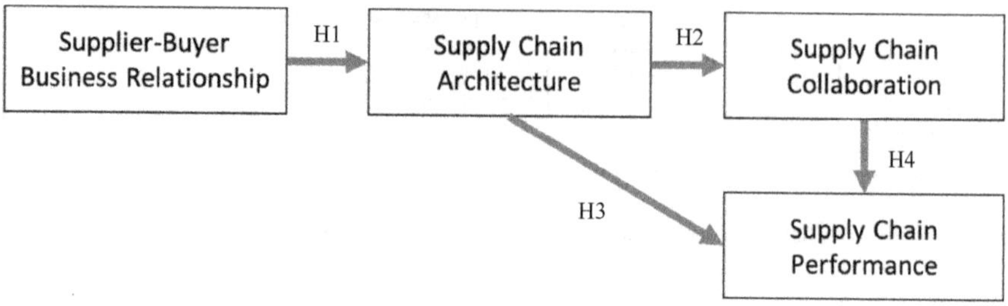

Fig. 9.3 Collaboration model between buyers and suppliers in the supply chain

The theoretical framework for the Indonesian pharmaceutical supply chain is depicted in Fig. 9.3.

9.3 Methodology

This study examines the level of pharmaceutical supply chain collaboration. It looks at the suitability of the supply chain coordination model proposed by Sheu et al. (2006) in the pharmaceutical industry in Indonesia. The research design is described further in the next sections.

9.3.1 Unit of Analysis, Population, and Research Sample

To address the research questions, the study surveyed pharmaceutical manufacturers in Indonesia. A manufacturing company is an appropriate unit of analysis because it is positioned relatively at the supply chain center. Furthermore, measuring the level of collaboration from the company side (towards suppliers and consumers) is more operationally feasible than measuring the supply chain network as a whole. Thus, the population of this study is pharmaceutical manufacturing companies (or those producing drugs) in Indonesia.

The Ministry of Health data shows that there were 214 manufacturing companies in Indonesia. Nevertheless, mainly 102 pharmaceutical companies in Indonesia were accessible, thus becoming the sampling frame. This study obtained 52 data from the pharmaceutical companies in Indonesia after approximately two months of data collection. Even though the data were limited, they represented 51% of the total sampling frame (the targeted population). Therefore, the data were considered sufficient for the study.

9.3.2 Data Collection Procedure

Albeit not too many pharmaceutical companies in Indonesia, access to companies in this industry is tricky because no directory lists profiles and addresses of all these companies. Therefore, the researchers tried to access companies through associations, namely GP-Pharmacy (Association of Indonesian Pharmaceutical Companies) and IAI (Indonesian Pharmacist Association). Both of these associations provided support and access to companies or practitioners who are the members of the association.

Given the obstacles faced in determining the sampling frame or gaining access to the company, random sampling is not possible. Instead, researchers contacted members assigned to GP-Pharmaceuticals and IAI and used a snowball method, where the respondents could suggest or invite colleagues working in other pharmaceutical manufacturers to participate in the survey.

9.3.3 Measurements

This measurement instrument was developed from previous research (Sheu et al., 2006; Yunus & Tadisina, 2016). The construct, measurement dimensions and question indicators can be seen in Table 9.1.

Table 9.1 Measurements and operational definition

No.	Construct	Dimension	Number of Items
1	Supplier–buyer business relationship *is the degree in which suppliers and retailers within the industry are connected*	Interdependence	4
		Intensity	5
		Trust	5
2	Supply chain architecture *is the extent of practices of designing and constructing relationships among supply chain members*	Information Sharing	7
		Inventory Systems	3
		Supply Chain Coordination Structure	3
		IT Capabilities	3
3	Supply chain collaboration *is the degree of partnership between suppliers and retailers within the industry*		4
4	Supply chain performance *is the achievement of results*		10

9.3.4 *Data Analysis*

This study performed a two-step testing as suggested by Anderson and Gerbing (1988). Firstly, we checked for the psychometric properties of the instrument. Secondly, we tested the hypotheses using the path analysis by JASP 0.14.1, open-source software suitable for statistical analysis.

9.4 Results

Table 9.2 shows the firms' profiles.

A two-step testing (Anderson & Gerbing, 1988) was employed for the full-scale survey data by assessing the instrument and evaluating the theoretical model. The descriptions and the results of each step are detailed in the following subsection.

9.4.1 *Assessment of the Measurement*

After data were collected, the data were used to test for the reliability and validity of the instruments. Before evaluating the measurement, we examined the data for

Table 9.2 The profiles of respondents/firms

		Frequency	%
Title	Firm owner	2	3.8%
	Senior manager, general manager or equivalent	11	21.2%
	Young managers or equivalent	19	36.5%
	Supervisor	15	28.8%
	Staff	5	9.6%
Tenure	< 3 years	21	40.4%
	3 to < 5 years	13	25.0%
	5 to < 7 years	9	17.3%
	7 to < 10 years	1	1.9%
	10 years or more	8	15.4%
Firm size	50–99 employees	5	9.6
	100–249 employees	6	11.5
	250–499 employees	17	32.7
	500–999 employees	9	17.3
	1000–4999 employees	6	11.5
	≥ 5000 employees	9	17.3

normality and multicollinearity issues. The normality assumption was checked using the Shapiro–Wilk test. All *p*-value of the Shapiro–Wilk showed significant value, which indicated that the data did not follow the normal distribution. However, we did not alter or modify the data as the Skewness statistics of all variables (Business Relationship, Architecture, Collaboration, and Performance) were still below the ±1.50. The results of the current test showed that all Tolerance values were above 0.20 (ranged from 0.667 to 0.752) and all VIF values were below 4.0 (ranged from 1.330 to 1.546) following the suggestion by Hair et al. (2006).

9.4.2 Hypotheses Testing

This study assessed the level of collaboration or synergy of the pharmaceutical industry in Indonesia. Based on the descriptive statistics as shown in Table 9.3, the Indonesian pharmaceutical firms are engaged in an above-average level of collaboration (3.92 of 5.00).

The results of the Exploratory Factor analysis were detailed in the Appendix. The Supplier–Buyer Business Relationship (BR), which conceptually comprised 3 dimensions, became a first-order construct (factor), so as the Supply Chain Architecture (AR). The Supply Chain Collaboration (CR) and Supply Chain Performance (PR) remained distinct first-order constructs. After a series of exploratory factor analyses, all items had above 0.5 loadings, which confirmed the validity of the items, and good reliability (above 0.7).

The goodness-of-fit of the model is presented in Table 9.4. The Chi-square test revealed an unsignificant result, indicated that the model fit the data. The goodness-of-fit indices are above 0.90, showing a good model (Gerbing & Anderson, 1988). The RMSEA, however, is above the 0.08 threshold, but we maintained the model since the RMR is low (below 0.05).

Table 9.5 shows the results of the hypothesis testing. All hypotheses were supported (p-value < 0.01), except for Hypothesis 3, which conjectured a direct effect of Supply Chain Architecture (AR) to Supply Chain Performance (PF). In other words, Supply Chain Collaboration (CL) fully mediated the relationship between AR and PF. The results are discussed further in the Discussion section.

Table 9.3 Descriptive statistics ($N = 52$)

	BR	AR	CL	PF
Mean	3.9775	3.7336	3.9231	3.8527
Std. Deviation	0.4329	0.5239	0.3656	0.3275

Legend: BR = Business Relationship; AR = Architecture; CL = Collaboration; PF = Performance

Table 9.4 Model goodness-of-fit indices

	Model
χ^2	3.0633
df	2.0000
χ^2 *p*-value	0.2162
Comparative Fit Index (CFI)	0.9715
Goodness of Fit Index (GFI)	0.9640
Tucker-Lewis Index (TLI)	0.9146
Bentler-Bonett Non-normed Fit Index (NNFI)	0.9146
Bentler-Bonett Normed Fit Index (NFI)	0.9294
Relative Noncentrality Index (RNI)	0.9715
Parsimony Goodness of Fit Index (PGFI)	0.8201
RMR	0.0089

Table 9.5 Results of the hypothesis testing

	Estimate	Std. Error	*p*	CI (lower)	CI (upper)	Conclusion
BR → AR	0.5453	0.1498	0.0003	0.2516	0.8389	*H1 supported*
AR → CL	0.3780	0.0814	0.0000	0.2185	0.5374	*H2 supported*
AR → PF	0.0801	0.0933	0.3903	−0.1027	0.2630	*H3 not supported*
CL → PF	0.3076	0.1337	0.0214	0.0456	0.5696	*H4 supported*

Legend: BR = Business Relationship; AR = Architecture; CL = Collaboration; PF = Performance

9.4.3 Non-Nested Model Comparison

To ensure that our empirically tested model was better than other potential (competing) model, this study performed another run for a model relating all exogenous variable (i.e., business relationship, architecture, and collaboration) to the endogenous variable (i.e., performance). Table 9.6 displays the statistical results essential for a model comparison testing, as suggested by Kline (2015).

Based on these results, we could conclude that the current model has a better fit, because its χ^2/df ratio, AIC, and BIC were lower than the competing model and thus better represented the data (Akaike, 1987; Kline, 2015).

Table 9.6 Results of model comparison testing

	χ^2/df	AIC	BIC
Current model	1.532	125.424	139.082
Competing model	3.023	127.383	142.993

9.5 Discussion

This study aims to investigate the extent of collaboration in the pharmaceutical industry in Indonesia. It conjectures that a higher level of collaboration would improve the supply chain performance, especially in terms of product availability and order fulfillment. This study further tests determinants of supply chain collaboration, namely business relationships and supply chain architecture. The implications of the results are discussed in the next subsection.

9.5.1 Implications for Theory

This study confirms the positive relationships among supply chain architecture, supply chain collaboration, and performance. As Yunus and Kurniawan (2015) argued, a lack of coordination between pharmaceutical manufacturers and their supply chain partners could result in un-sync logistical executions and hence poor supply chain performance. This study provides empirical evidence of the positive impact of supply chain architecture on collaboration, and further on performance. In this study, collaboration fully mediates the relationship between supply chain architecture and performance.

Based on the findings, this study also corroborates the supplier-buyer business relationship as the determinant of supply chain architecture. This is aligned with previous studies (Collier & Sarkis, 2021; Saeed et al., 2019), which argue that an increase in interdependence, intensity, and trust would form superior inventory systems, information sharing practice, and information technology capability as dimensions of supply chain architecture.

Supply chain performance is critical in the pharmaceutical industry due to its significant and direct impact on society. This study measures the supply chain performance through internal impact and outward-orientation results, such as on-time delivery to customers, order fulfilment, and service excellence. This is aligned with insights from Narayana et al. (2014), who observed a shift from the company's internal focus to the supply chain network to the retail level (healthcare services). As the manufacturers improve their operational performance, people can appreciate the results through available drugs at affordable prices, which has been challenging to achieve thus far in developing countries.

9.5.2 Implications for Practice

In Indonesia, there are many issues related to the distribution of pharmaceutical products that result in people being unable to obtain medicines. This situation happens not only in remote areas but, ironically, also in big cities in Indonesia. This study confirmed that collaboration among supply chain entities within the pharmaceutical industry—namely, suppliers, manufacturers, wholesalers, and retailers—would improve supply chain performance regarding inventory availability, order fulfillment, and return.

Collaboration indicates that supply chain partners perform integrated logistical activities by planning and engaging in intensive coordination. This study suggests that managers improve the business relationship by setting up communication channels, appointing a person in charge from each party, and establishing systems to monitor the supply chain processes. As the business relationship grows, companies would improve their supply chains architecture, such as electronic data interchange and information sharing. Thus, collaboration would also increase, and the Indonesian pharmaceutical industry would obtain a better supply chain performance.

9.5.3 Post-COVID Implications for Supply Chains in Pharmaceutical Company

When the World Health Organization (WHO) received a report of a new pneumonia case of unknown cause on December 31, 2019, and WHO declared a COVID-19 pandemic on March 11, 2021, no one realized how big the disaster was. It turns out, even after more than 20 months of us 'living life amid a pandemic', this phenomenon is not over. This coronavirus pathogen moves quickly to spread throughout the world and kills tens of millions of people, shakes the economy of almost all countries in the world, and disrupts national stability. The acceleration of the outbreak is triggered by climate change, urbanization, as well as lack of water and poor sanitation (World Bank, 2020, 2021b). Many countries are now facing a more substantial second wave. Some countries even experienced a third wave, such as the UK, Germany, Brazil, and countries in Africa. This pandemic is far from over.

On the other hand, the pandemic leads us to further consequences. During the pandemic, consumers use online channels for various activities previously done physically. Business, respond. In Indonesia, real growth is taking place. The e-commerce business is proliferating. This sector is projected to increase by 37.4% from 2020, or to Rp351.1 trillion. The compound annual growth rate (CAGR) is estimated at 19.2% between 2020 and 2024, bringing it to Rp707.6 trillion by 2024 (Global Data, 2021).

Moreover, as reported at the official Indonesian Information Portal (Indonesia.go.id), the accumulated value of purchases through e-commerce sites or apps (gross merchandise value/GMV) rose 54% from USD21 billion in 2019 to USD32 billion. This value is estimated to increase to USD83 billion in 2025. In line with the increase in the e-commerce business, digital banking transactions are projected to increase to around Rp32,206 trillion in 2021 or grow 19.1% from the realization of 2020 transactions (Hidranto, 2021).

The COVID-19 pandemic has undoubtedly brought us to the Industrial Revolution 4.0 through the acceleration of digitalization. A report from the McKinsey Global Survey, which looked at 899 senior directors and managers (their geographic origin is not specified), shows that adoption of digitization is much faster during a pandemic. The Asia–Pacific region has experienced a higher acceleration of digitalization of the interaction-with-consumer process than the Americas, Europe, and Global. The level of digital transformation of all/part of the goods or services offered in the Asia–Pacific region is the highest compared to other regions, which is experiencing an average acceleration of more than ten years, especially in the health and pharmaceutical sectors, financial services, and professional services (McKinsey, 2020).

The pandemic has made organizations realize the weaknesses of their supply chain management and begin to make improvements (Shih, 2020). When imported raw materials are not possible, the company starts to procure locally, expand the supply chain network, and develop new vendors (Shih, 2020; Harapko, 2021). Drones are now being used to deliver vaccines, blood, and medicines to various regions to reduce human contact (*The Economist*, 2021).

The pharmaceutical industry has felt the real impact of weak coordination in the supply chain of drugs, medical devices, and COVID-19 vaccines. When the second pandemic wave hit Indonesia, the public felt the unprepared supply of medicines and vaccines. However, the Indonesian government swiftly coordinated pharmaceutical manufacturers, distributors, hospitals, and oxygen and oxygen cylinders producers. All parties must move in the same rhythm to achieve optimal supply chain performance. This situation assures us about the importance of leadership in driving collaboration in supply chain networks, as argued by Yunus and Kurniawan (2015), as well as the essential factor of cooperation and coordination within the pharmaceutical supply chain.

9.6 Conclusion

This study examines the level of collaboration along the pharmaceutical supply chain in Indonesia. From the study results, the level of cooperation between pharmaceutical companies and their distributors is considerably good (3.92 out of 5.00) but still needs to be improved to the retail level, namely pharmacies, drug stores, and public health services. This study shows that the intensity and level of trust in building business relationships between suppliers and buyers is one of the drivers of improving supply chain architecture towards higher collaboration. Amid the current heavy pandemic, partnership and coordination between various parties in increasing the accessibility of drugs and vaccines are the main determinants of controlling COVID-19 virus contamination.

Appendix

Loadings and Reliability (Cronbach's Alpha)

Business Relationship (BR)—0.812		
Interdep1	Our company discusses the principal-distributor relationship at the strategic planning level	0.689
Interdep2	Our company holds periodic principal-distributor meetings regarding target agreements and supply chain performance	0.861
Interdep3	With regard to the procurement process, we as principals and distributors have an interconnected system to integrate supply information	0.718
Interdep4	In deciding something regarding the procurement of goods, we usually discuss first with the distributor	0.506
Intensity3	We develop professional procurement personnel who are directed to achieve a competitive strategy	0.617
Intensity5	Top Management pays full attention to developing and maintaining supply chain HR loyalty (develop and retain employee engagement)	0.537

Architecture (AR)—0.883		
InfSharing2	We exchange procurement information via the internet	0.778
InfSharing6	Our company shares available inventory information with distributors	0.480
InfSharing7	Distributors share existing inventory level information with us	0.543
Inventory1	Inventory management in our company is carried out in an integrated and computerized manner between the principal and distributor	0.653
IT1	Routine communication between principal and distributor is computerized	0.840
IT2	Principals and distributors use technology that enables electronic transaction processing	0.919
IT3	Coordination between companies can be achieved using electronic links	0.842

Collaboration (CL)—0.790		
CL1	We carry out logistics activities with distributors with coordination that makes it easier for each other	0.554
CL2	We carry out logistics activities with well-integrated distributors	0.987
CL3	Principal and distributor logistics integration includes distribution, transportation, and/or warehousing activities	0.600

(continued)

(continued)

Collaboration (CL)—0.790		
CL4	The flow of information and goods runs smoothly and is integrated between principals and distributors	0.666

Performance (PF)—0.788		
PF1	Our company has a good record of on-time delivery	0.7395
PF2	Our company is able to fulfill customer orders based on information from distributors quickly	0.6449
PF3	Our company provides excellent service to distributors	0.6164
PF4	Sales of our company's products are experiencing good growth	0.6459
PF5	Low company product sales returns	*deleted*
PF6	Our company's product sales returns are decreasing	0.7511
PF7	Our company profit is growing	0.9091
PF8	The market share of our company's products has increased	0.5107
PF9	My company's return on investment has increased	0.8939
PF10	Our company's investment development is getting higher	*deleted*

References

Akaike, H. (1987). Factor analysis and AIC. *Psychometrika, 52*(3), 317–332.

Al-Doori, J. A. (2019). The impact of supply chain collaboration on performance in automotive industry: Empirical evidence. *Journal of Industrial Engineering and Management, 12*(2), 241–253.

Anderson, J. C., & Gerbing, D. W. (1988). Structural equation modeling in practice: A review and recommended two-step approach. *Psychological Bulletin, 103*(3), 411–423.

Chaudhuri, A., Boer, H., & Taran, Y. (2018). Supply chain integration, risk management and manufacturing flexibility. *International Journal of Operations & Production Management.*

Chen, L., Zhao, X., Tang, O., Price, L., Zhang, S., Zhu, W. (2017). Supply chain collaboration for sustainability: A literature review and future research agenda. *International Journal of Production Economics.*

Collier, Z. A., & Sarkis, J. (2021). The zero trust supply chain: Managing supply chain risk in the absence of trust. *International Journal of Production Research, 59*(11), 3430–3445.

Danese, P., Romano, P., & Formentini, M. (2013). The impact of supply chain integration on responsiveness: The moderating effect of using an international supplier network. *Transportation Research Part E: Logistics and Transportation Review, 49*(1), 125–140.

The Economist. (2021). *How covid-19 is boosting innovation.* https://www.economist.com/films/2021/03/10/how-covid-19-is-boosting-innovation

Flynn, B. B., Huo, B., & Zhao, X. (2010). The impact of supply chain integration on performance: A contingency and configuration approach. *Journal of Operations Management, 28*(1), 58–71.

Gerbing, D. W., & Anderson, J. C. (1988). An updated paradigm for scale development incorporating unidimensionality and its assessment. *Journal of Marine Research, 25*(2), 186–192.
Global Data. (2021). *Indonesia's e-commerce market continues to surge amid COVID-19 pandemic.* https://www.globaldata.com/indonesias-e-commerce-market-continues-surge-amid-covid-19-pandemic-says-globaldata/
Ha, B. C., Park, Y. K., & Cho, S. (2011). Suppliers' affective trust and trust in competency in buyers its effect on collaboration and logistics efficiency. *International Journal of Operations & Production Management, 31*(1–2), 56–77. https://doi.org/10.1108/01443571111098744
Hair, J. F. J., Black, W. C., Babin, B. J., Anderson, R. E., & Tatham, R. L. (2006). *Multivariate data analysis.* Prentice Hall.
Harapko, S. (2021). *How COVID-19 impacted supply chains and what comes next.* https://www.ey.com/en_id/supply-chain/how-covid-19-impacted-supply-chains-and-what-comes-next
Hidranto, F. (2021). *Bisnis e-commerce semakin gurih.* https://www.indonesia.go.id/kategori/indonesia-dalam-angka/2534/bisnis-e-commerce-semakin-gurih
Kementerian Kesehatan RI. (2015). *Rencana Strategis Kementerian Kesehatan (2015–2019).* Kementerian Kesehatan RI.
Kiely, D. (2004). The state of pharmaceutical industry supply planning and demand forecasting. *The Journal of Business Forecasting Methods & Systems, 23*(3), 20–22.
Kline, R. B. (2015). *Principles and practice of structural equation modeling.* Guilford Publications.
Lambert, D. M., & Cooper, M. C. (2000). Issues in supply chain management. *Industrial Marketing Management, 29*(1), 65–83.
Lee, C., & Ha, B. C. (2020). The impact of interactional justice and supply-chain collaboration on sustainable SCM performance: The case of multinational pharmaceutical firms. *The Journal of Asian Finance, Economics and Business, 7*(2), 237–247.
McKinsey. (2020). *How COVID-19 has pushed companies over the technology tipping point—And transformed business forever.* https://www.mckinsey.com/business-functions/strategy-and-corporate-finance/our-insights/how-covid-19-has-pushed-companies-over-the-technology-tipping-point-and-transformed-business-forever
Moosivand, A., Ghatari, A. R., & Rasekh, H. R. (2019). Supply chain challenges in pharmaceutical manufacturing companies: Using qualitative system dynamics methodology. *Iranian Journal of Pharmaceutical Research, 18*(2), 1103.
Munir, M., Jajja, M. S. S., Chatha, K. A., & Farooq, S. (2020). Supply chain risk management and operational performance: The enabling role of supply chain integration. *International Journal of Production Economics, 227*, 107667.
Mustamu, R. H. (2007). Manajemen rantai pasokan industri farmasi di Indonesia. *Jurnal Manajemen Dan Kewirausahaan (or J Man Entr), 9*(2), 99.
Narayana, S. A., Pati, R. K., & Vrat, P. (2014). Managerial research on the pharmaceutical supply chain—A critical review and some insights for future directions. *Journal of Purchasing & Supply Management, 20*(1), 18–40.
OECD. (2018). *Excessive pricing in pharmaceutical markets—Note by Indonesia.* https://one.oecd.org/document/DAF/COMP/WD(2018)114/en/pdf
Ralston, P. M., Richey, R. G., & Grawe, S. J. (2017). The past and future of supply chain collaboration: A literature synthesis and call for research. *The International Journal of Logistics Management, 28*(2).
Saeed, K. A., Malhotra, M. K., & Abdinnour, S. (2019). How supply chain architecture and product architecture impact firm performance: An empirical examination. *Journal of Purchasing & Supply Management, 25*(1), 40–52.
Sheu, C., Yen, H. R., & Chae, B. (2006). Determinants of supplier-retailer collaboration: Evidence from an international study. *International Journal of Operations & Production Management, 26*(1), 24–49.
Shih, W. C. (2020). Global supply chains in a post-pandemic world. *Harvard Business Review, 98*(5), 82–89.

Tarifa-Fernandez, J., & De Burgos-Jiménez, J. (2017). Supply chain integration and performance relationship: A moderating effects review. *The International Journal of Logistics Management*.

Wang, M., & Jie, F. (2019). Managing supply chain uncertainty and risk in the pharmaceutical industry. *Health Services Management Research, 33*(3), 156–164.

World Bank. (2020). *The global economic outlook during the COVID-19 pandemic: A changed world*. https://www.worldbank.org/en/news/feature/2020/06/08/the-global-economic-outlook-during-the-covid-19-pandemic-a-changed-world

World Bank. (2021a). *World development indicators*. http://wdi.worldbank.org/

World Bank. (2021b). *Pandemic preparedness and COVID-19 (coronavirus)*. https://www.worldbank.org/en/topic/pandemics

Yu, X., Li, C., Shi, Y., & Yu, M. (2010). Pharmaceutical supply chain in China: Current issues and implications for health system reform. *Health Policy, 97*(1), 8–15.

Yunus, E. N., & Kurniawan, T. (2015). Revealing unsuccessful collaboration: A case of buyer-supplier relationship in the pharmaceutical industry. *Supply Chain Forum: An International Journal, 16*(2), 14–28.

Yunus, E. N., & Tadisina, S. K. (2016). Drivers of supply chain integration and the role of organizational culture: Empirical evidence from Indonesia. *Business Process Management Journal, 22*(1), 89–115.

Thematic Bibliography

The learned scholars who contributed with their chapters have used a large body of existing literature on the various relevant topics and themes of supply chain management. To facilitate future researchers and scholars in the rich domain of the supply chain, we list those references, 500 in total, as a thematic bibliography with two major themes, which are (i) Understanding Supply Chain Dynamics—Theoretical Perspectives, and (ii) Understanding Supply Chain Dynamics—Modeling-Based Empirical Solutions. Geographically, the studies listed here cover the regions and countries of the world. The researchers and practitioners especially from the supply chain domain can avail this list. In particular, while building this thematic bibliography, I had our graduate students in my mind and here are provided with a set of relevant references as a starting point of their literature review journey.

Theme 1: Understanding Supply Chain Dynamics—Theoretical Perspectives

A total of 168 studies covering both theoretical and empirical including the state-of-the reviews relevant to various topics of this theme are included here. Key topics addressed in these studies include Global supply chains, Supply chain resilience, COVID-19 crisis, Supply chain risk assessment, Blockchain-based supply chain, Supply chain disruption, Social network analysis, Product development, Buyer and supplier relationships, Decentralization of the firm, Specialized supplier networks, Strategic partnerships, Strategic supplier selection, Impact of the Covid-19 pandemic, Supply chain and human resource management practices, Supply chain risk management, Forecasting and planning, Global supply chain resilience, Mitigation and contingency strategies for managing supply chains.

H. Qudrat-Ullah (ed.), *Understanding the Dynamics of New Normal for Supply Chains*, Understanding Complex Systems,
https://doi.org/10.1007/978-3-031-07333-5

References for Theme 1

AFP News. (2020, January 29). *China Virus Injects Fears for Global Supply Chains*. Retrieved from https://www.ibtimes.com/china-virus-injects-fears-global-supply-chains-2912001

Ali, M. H., Suleiman, N., Khalid, N., Tan, K. H., Tseng, M. L., & Kumar, M. (2021). Supply chain resilience reactive strategies for food SMEs in coping to COVID-19 crisis. *Trends in Food Science & Technology, 109*, 94–102.

Ambulkar, S., Blackhurst, J., & Grawe, S. (2015). Firm's resilience to supply chain disruptions: Scale development and empirical examination *Journal of Operations Management, 33*(34), 111–122.

AMD Consulting. (2018, January 5). *Inventory optimization through supply chain risk assessment*. https://www.amdconsulting.com/inventory-optimization-supply-chain-risk-assessment/. Accessed 23 Feb 2022.

Azzi, R., Chamoun, R. K., & Sokhn, M. (2019). The power of a blockchain-based supply chain. *Computers & Industrial Engineering, 135*, 582–592.

Balachandran, K. (2005). *Quality implications of warranties in a supply chain*. Retrieved from http://mansci.journal.informs.org/content/51/8/1266.full.pdf. Accessed 13 Feb 2022.

Barbieri, P., Bofelli, A., Elia, S., Fratocchi, L., Kalchschmidt, M., & Samson, D. (2020). What can we learn about reshoring after Covid-19? *Operations Management Research, 13*(3), 131–136.

Baxter, D. (n.d.). Supply chain disruption: The bad, the ugly, & the future, real-time visibility—Supply chain 24/7. Retrieved from https://www.supplychain247.com/article/supply_chain_disruption_the_bad_the_ugly_the_future

Bell, J. (2020, November 16). Who are the world's biggest needle and syringe manufacturers? *NS Medical Devices*. www.nsmedicaldevices.com/analysis/biggest-needle-syringe-manufacturers-companies/

Bendapudi, N., & Leone, R. P. (2002). Managing business-to-business customer relationships following key contact employee turnover in a vendor firm. *Journal of Marketing, 66*(2), 83–101.

Benton, M. C., Radziwill, N. M., Purritano, A. W., & Gerhart, C. J. (2018). Blockchain for supply chain: Improving transparency and efficiency simultaneously. *Software Quality Professional, 20*(3).

Borgatti, S. P., & Li, X. (2009). On social network analysis in a supply chain context. *Journal of Supply Chain Management, 45*(2), 5–22.

Bosch CEO warns coronavirus could hit global auto supply chains. (2020, January 29). Retrieved from https://www.msn.com/en-us/finance/companies/bosch-ceo-warns-coronavirus-could-hit-global-auto-supply-chains/ar-BBZqGvD

Brinken, J., Pabsch, C., & Behrendt, F. (2022). Decarbonization potential of logistic 4.0 technologies in apple supply chains. *Procedia Computer Science, 200*, 461–470.

Brix-Asala, C., Geisbüsch, A. K., Sauer, P. C., Schöpflin, P., & Zehendner, A. (2018). Sustainability tensions in supply chains: A case study of paradoxes and their management. *Sustainability, 10*(2), 424–433.

Brown, S. L., & Eisenhardt, K. M. (1995). Product development: Past research, present findings, and future directions. *Academy of Management Review, 20*(1), 343–378.

Brunnermeie, S. B., & Martin, S. A. (2002) Interoperability costs in the US automotive supply chain. *Supply Chain Management: An International Journal, 7*(2), 71–82.

Brynjolfsson, E., Horton, J. J., Ozimek, A., Rock, D., Sharma, G., & TuYe, H. Y. (2020). *COVID-19 and remote work: An early look at US data* (No. w27344). National Bureau of Economic Research.

Butt, A. S. (2021). Strategies to mitigate the impact of COVID-19 on supply chain disruptions: A multiple case analysis of buyers and distributors. *The International Journal of Logistics Management*, Vol. ahead-of-print, No. ahead-of-print. https://doi.org/10.1108/IJLM-11-2020-0455

Calcaterra, C., & Kaal, W. A. (2020). Reputation protocol for the internet of trust. In *Legal tech and the new sharing economy* (pp. 155–179). Springer.

Callikan, A. Coronavirus/COVID-19 worldwide cases live data & statistics. *Covid Statistics*. http://covidstatistics.org/

Carter, J. R., Smeltzer, L., & Narasimhan, R. (1998). The role of buyer and supplier relationships in integrating TQM through the supply chain. *European Journal of Purchasing and Supply Management, 4*(4), 223–234.

Chao, G. (2009). *Quality improvement incentives and product recall cost-sharing contracts.* Retrieved from http://mansci.journal.informs.org/content/55/7/1122.full.pdf. Accessed 23 Feb 2022.

Chopra, S., & Sodhi, M. S. (2014). Reducing the risk of supply chain disruptions. *MIT Sloan Management Review, 55*(3, Spring), 72.

Christie, A. A., Joye, M. P., & Watts, R. L. (2003). Decentralization of the firm: Theory and evidence. *Journal of Corporate Finance, 9*(1), 3–36.

Coley, L. S., Lindemann, E., & Wagner, S. M. Effects of suppliers' reputation on the future of buyer-supplier relationships: The mediating roles of outcome fairness and trust. *Journal of Supply Chain Management, 47*(2), 29–48.

Considering Relationships in Supply Chain Management. *Engineering Management*. IEEE Transactions on 48, 307–318. https://doi.org/10.1109/17.946529

Coronavirus (COVID-19) Origin: Cause and how it spreads. *Medical News Today*. MediLexicon International. www.medicalnewstoday.com/articles/coronavirus-causes

Cusumano, M., & Takeishi, A. (1991). Supplier relations and management: a survey of Japanese, Japanese transplant, and US auto plants. *Strategic Management Journal, 12*(6), 563–588.

Daft, R. L., & Lengel, R. H. (1986). Organizational information requirements, media richness and structural design. *Management Science,* 32(5), 554–571.

Deepankar Sinha Indian Institute of Foreign Trade, Kolkata, West Bengal, India. (2020, November 4). *Virupaxi Bagodi: The supply chain disruption framework post COVID-19: A system dynamics model*. Research Article. https://doi.org/10.1177/0015732520947904

Desai, A., Nguyen, H. N., & Nagda, B. R. A. (2021). Honoring culture, holding complexity: Synthesis and emerging possibilities in dialogue. In *Global perspectives on dialogue in the classroom* (pp. 175–199). Palgrave Macmillan.

Doran, D. (2004). Rethinking the supply chain: An automotive perspective. *Supply Chain Management: An International Journal, 9*(1), 102–109.

Dowty, R. A., & Wallace, W. A. (2010). Implications of organizational culture for supply chain disruption and restoration. *International Journal of Production Economics, 126*(1), 57–65. ISSN 0925-5273. https://doi.org/10.1016/j.ijpe.2009.10.024, https://www.sciencedirect.com/science/article/pii/S0925527309003934)

Dyer, J. H. (1996). Specialized supplier networks as a source of competitive advantage: Evidence from the auto industry. *Strategic Management Journal, 17*(3), 271–291.

Eden, C., Ackermann, F., & Vito, V. (2021, June). Improvisation and emergent strategizing: The role of group support systems. In *International Conference on Group Decision and Negotiation* (pp. 16–24). Springer.

Ekanayake, M. A. C., Shen, G. Q. P., M. M. Kumaraswamy. (2021). Identifying supply chain capabilities of construction firms in industrialized construction. *Production Planning & Control, 32*(4), 303–321.

Ellram, L. M. (1990). The supplier selection decision in strategic partnerships. *Journal of Purchasing and Materials Management, 20*(4), 8–14.

Evangelista, P., Mogre, R., Perego, A., Raspagliesi, A., & Sweeney, E. (2012). A survey based analysis of IT adoption and 3PLs' performance. *Supply Chain Management: An International Journal, 17*(2), 172–186.

Florez-Lopez, R. (2007). Strategic supplier selection in the added-value perspective: A CI approach. *Information Sciences, 177*(5), 1169–1179.

Flynn, B., Cantor, D., Pagell, M., Dooley, K. J., & Azadegan, A. (2021). From the editors: Introduction to managing supply chains beyond Covid-19—Preparing for the next global mega-disruption. *Journal of Supply Chain Management, 57*(1), 1–36.

Foss, N. J. (2020). The impact of the Covid-19 pandemic on firms' organizational designs. *Journal of Management Studies*.

Francisco, K., & Swanson, D. (2018). The supply chain has no clothes: Technology adoption of blockchain for supply chain transparency. *Logistics, 2*(1), 2–11.

Franklin, S., & Graesser, A. (1996, August). Is it an Agent, or just a Program?: A taxonomy for autonomous agents. In *International workshop on agent theories, architectures, and languages* (pp. 21–35). Springer.

Frederico, G. F. (2021). Towards a supply chain 4.0 on the post-COVID-19 pandemic: A conceptual and strategic discussion for more resilient supply chains. *Rajagiri Management Journal*, Vol. ahead-of-print, No. ahead-of-print. https://doi.org/10.1108/ramj-08-2020-0047

Friday, D., et al. (2021). A collaborative approach to maintaining optimal inventory and mitigating stockout risks during a pandemic: Capabilities for enabling health-care supply chain resilience. *Journal of Humanitarian Logistics and Supply Chain Management*, Vol. ahead-of-print, No. ahead-of-print. https://doi.org/10.1108/jhlscm-07-2020-0061

Fu, H., Zhao, C., Cheng, C., & Ma, H. (2020). Blockchain-based agri-food supply chain management: Case study in China. *International Food and Agribusiness Management Review, 23*(5), 667–679.

Gorchels, L. (2004). *The manager's guide to distribution channels*. McGraw-Hill.

Gray, D., & Vander Wal, T. (2014). *The connected company*. O'Reilly Media, Inc.

Gray, S. (2019, May 10). *Can your business handle these supply chain disruptions?* Retrieved from https://www.thebalancesmb.com/preparing-for-supply-chain-disruptions-4589604

Gurman, M., Wu, D., & Bloomberg. (2020, January 28). *Apple supply chain braces for disruption from coronavirus*. Retrieved from https://www.sfgate.com/business/article/Apple-supply-chain-braces-for-disruption-from-15010387.php. Accessed 12 Mar 2022.

Haeckel, S. H. (1995). Adaptive enterprise design: The sense-and-respond model. *Planning Review*.

Hales, B. M., & Pronovost, P. J. (2006). The checklist—A tool for error management and performance improvement. *Journal of Critical Care, 21*(3), 231–235.

Hedwall, M. (2020, June 22). The Ongoing Impact of COVID-19 on global supply chains. *World Economic Forum*. www.weforum.org/agenda/2020/06/ongoing-impact-covid-19-global-supply-chains/

Helper, S., & Sako, M. (1995). Supplier relations in Japan and the United States: Are they converging. *Sloan Management Review, 36*(3), 77–84.

Heun, B. (2020, February 11). *Protecting your balance sheet against supply chain risks, disruptions*. Retrieved from https://www.bizjournals.com/philadelphia/news/2020/02/11/protecting-your-balance-sheet-against-supply-chain.html

Hill, A., & Hill, T. (2009). Manufacturing operations strategy (3rd ed.). Palgrave Macmillan.

Hill, T. (2000). *Manufacturing strategy: Text and cases* (3rd ed.). Irwin McGraw-Hill.

Hines, T. (2004). *Supply chain strategies: Customer driven and customer focused*. Elsevier.

Hiremath, P., Kowshik, C. S., Manjunath, M., & Shettar, M. (2020). COVID 19: Impact of lock-down on mental health and tips to overcome. *Asian Journal of Psychiatry, 51*, 102088.

Hitt, M. A., Holmes, R. M., Jr., & Arregle, J. L. (2021). The (COVID-19) pandemic and the new world (dis)order. *Journal of World Business, 56*(4), 101210. https://scm.ncsu.edu/scm-articles/article/supply-chain-disruptions

IndustryStar. (2018, September 9). *5 supply chain disruptions & mitigation tactics*. Retrieved from https://www.industrystarsolutions.com/blog/2018/09/5-supply-chain-disruptions-reduce-impact/

Ivanov, D., Dolgui, A., Sokolov, B., & Ivanova, M. (2017). Literature review on disruption recovery in the supply chain. *International Journal of Production Research, 55*(20), 6158–6174.

Jia, W., & Zhou, W. (2004). *Distributed network systems: From concepts to implementations* (Vol. 15). Springer Science & Business Media.

Johnson, B. (1993). Polarity management. *Executive Development, 6*, 28–28.

Kannan, V. R., & Tan, K. C. (2013). Supplier selection and assessment: Their impact on business performance. *Journal of Supply Chain Management, 38*(3), 11–21.

Kaplan, R. (1998). *Cost and effect: Using integrated cost system to drive profitability and performance* (p. 13). Harvard Business School Press.

Karl, K. A., Peluchette, J. V., & Aghakhani, N. (2021). Virtual work meetings during the COVID-19 pandemic: The good, bad, and ugly. *Small Group Research*. http://doi.org/10.1177/10464964211015286

Khudhair, H. Y., Alsaud, A. B., Alsharm, A., Alkaabi, A., & AlAdeedi, A. (2020). The impact of COVID-19 on supply chain and human resource management practices and future marketing. *International Journal of Supply Chain Management, 9*(5), 1681.

Kiers, J., Seinhorst, J., Zwanenburg, M., & Stek, K. (2022). Which Strategies and corresponding competencies are needed to improve supply chain resilience: A COVID-19 based review. *Logistics, 6*(12), 1–17.

Kim, D. H. (1995). Systems archetypes as dynamic theories. *The Systems Thinker, 6*(5), 6–9.

Kirkley, J. (2012, March 6). *Supply chain risk management copes with evolving threats*. Retrieved from https://www.enterpriseai.news/2012/03/06/coping_with_evolving_threats_with_supply_chain_risk_management/

Klimoski, R., & Mohammed, S. (1994). Team mental model—Construct or metaphor. *Journal of Management, 20*(2), 403–437.

Komatsu Cipriani, T., Deserti, A., Kleverbeck, M., Rizzo, F., & Terstriep, J. (2020). Business models & social innovation: Mission-driven versus profit-driven organisations. *International Review of Applied Economics, 34*(5), 541–566.

Kouvelis, P., et al. (2011). *Handbook of integrated risk management in global supply chains*. Wiley, Incorporated. ProQuest Ebook Central, http://ebookcentral.proquest.com/lib/asulib-ebooks/detail.action?docID=697522. Created from asulib-ebooks on 2020-06-19 11:07:23.

Krause, D. R. (1997). Supplier development: Current practices and outcomes. *International Journal of Purchasing and Materials Management, 33*(2), 12–19.

Kumar, B., & Sharma, A. (2021). Managing the supply chain during disruptions: Developing a framework for decision-making. *Industrial Marketing Management, 97*, 159–172.

Kumar, S., Liu, J., & Scutella, J. (2015). The impact of supply chain disruptions on stockholder wealth in India. *International Journal of Physical Distribution & Logistics Management, 45*(9/10), 938–958.

Kunovjanek, M., & Wankmüller, C. (2020). An analysis of the global additive manufacturing response to the COVID-19 pandemic. *Journal of Manufacturing Technology Management, 32*(9), 75–100. https://doi.org/10.1108/jmtm-07-2020-0263

Kurokawa, S., Manabe, S., & Rassameethes, B. (2008). *Journal of Organizational Computing and Electronic Commerce, 18*(1), 1–33.

Kurtz, C. F., & Snowden, D. J. (2003). The new dynamics of strategy: Sense-making in a complex and complicated world. *IBM Systems Journal, 42*(3), 462–483.

Lakonishok, J., Shleifer, A., & Vishny, R. W. (1994). Contrarian investment, extrapolation, and risk. *The Journal of Finance, 49*(5), 1541–1578.

Lam, K.-C., Tao, R., & Lam, M. C.-K. (2010). A material supplier selection model for property developers using fuzzy principal component analysis. *Automation in Construction, 19*(5), 608–618.

Lee, M. (2003). A high-quality-supplier selection model for supply chain management and iso 9001. *Systems, 14*(3), 225–232.

Li, S., Ragu-Nathan, B., Ragu-Nathan, T. S., & Subba Rao, S. (2006). The impact of supply chain management practices on competitive advantage and organizational performance. *Omega, 34*(2), 107–124.

Lijuan, M. (2002). The primary study of supplier selection based on supply chain management. *Industrial Engineering Management, 6*, 23–25.

Liu, J., Sarkar, S., Kumar, S., & Jin, Z. (12/2016). An analysis of stock market impact from supply chain disruptions in Japan. *International Journal of Productivity and Performance Management*.

Lohmer, J., Petzok, L., & Lasch, R. (2021). Governance design of blockchain consortia for efficient and transparent procurement and supply chain management. In *Supply Management Research* (pp. 117–134). Springer Gabler.

Lorentz, H., & Hilmola, O.-P. (10/2012). Confidence and supply chain disruptions. *Journal of Modelling in Management, 7*(3), 328–356.

Lummus, R. R., & Vokurka, R. J. (1999). Defining supply chain management: A historical perspective and practical guidelines. *Industrial Management & Data Systems, 99*(1), 11–17.

Luthy, J. (2018, September 8). *5 major supply chain disruptions and how to reduce their impact.* Retrieved from https://www.industrystarsolutions.com/blog/category/supply-chain-disruption/

Matson, J. E., & Matson, J. O. (2007). Just-in-time implementation issues among automotive suppliers in the southern USA. *Supply Chain Management: An International Journal, 12*(6), 432–443.

Mazzei, D., Baldi, G., Fantoni, G., Montelisciani, G., Pitasi, A., Ricci, L., & Rizzello, L. (2020). A blockchain tokenizer for industrial IOT trustless applications. *Future Generation Computer Systems, 105*, 432–445.

Mentzer, J. T., et al. (2001). Defining supply chain management. *Journal of Business Logistics, 22*(2), 1–25.

Monczka, R. M., Peterson, K. J., Handfield, R. B., & Ragatz, G. L. (1998). Success factors in strategic supplier alliances: The buying company perspective. *Decision Sciences, 29*(1), 69–73.

Montgomery, O. (2020, October 2). 5 types of supply chain disruption with COVID-19 examples. *Software Advice.* www.softwareadvice.com/resources/supply-chain-disruption-types/

Movahedi, B., Lavassani, K., & Kumar, V. (2009). Transition to B2B e-marketplace enabled supply chain: Readiness assessment and success factors. *The International Journal of Technology, Knowledge and Society, 5*(3), 75–88.

Muller, R. (2010). *Supplier relationship management (SRM): Basic concepts, strategies, potential.* GRIN Verlag.

Nadini, M., Alessandretti, L., Di Giacinto, F., Martino, M., Aiello, L. M., & Baronchelli, A. (2021). Mapping the NFT revolution: Market trends, trade networks, and visual features. *Scientific Reports, 11*(1), 1–11.

NC State University. (2004, September 28). *Supply chain disruptions.*

Nikolopoulos, K., Punia, S., Schäfers, A., & Tsinopoulos, C., & Vasilakis, C. (2021). Forecasting and planning during a pandemic: COVID-19 growth rates, supply chain disruptions, and governmental decisions. *European Journal of Operational Research, 290*(1), 99–115. ISSN 0377-2217. https://doi.org/10.1016/j.ejor.2020.08.001, https://www.sciencedirect.com/science/article/pii/S0377221720306913

Nyce, C., & Cpcu, A. (2007). Predictive analytics white paper. *American Institute for CPCU: Insurance Institute of America*, 9–10.

O'Brien, J. (2014). *Supplier relationship management: Unlocking the hidden value in your supply base.* Kogan Page.

O'Sullivan, S. (2019, August 14). *Supply chain disruption: Aligning business strategy and supply chain T*. Retrieved from https://www.koganpage.com/article/supply-chain-disruption-aligning-business-strategy-and-supply-chain-tactic

Obrenovic, B., Du, J., Godinic, D., Tsoy, D., Khan, M. A. S., & Jakhongirov, I. (2020). Sustaining enterprise operations and productivity during the COVID-19 pandemic: Enterprise effectiveness and sustainability model. *Sustainability, 12*(15), 5981.

Oosterhuis, M., Vaart, T. v. d., & Molleman, E. (2012). The value of upstream recognition of goals in supply chains. *Supply Chain Management: An International Journal, 17*(6), 582–595.

Ovalle, O. R., & Marquez, A. C. (2003). The effectiveness of using e-collaboration tools in the supply chain: An assessment study with systems dynamics. *Journal of Purchasing and Supply Management, 9*(4), 151–163.

Ozgur, C. (2020, January 28). *Coronavirus supply chain disruption will affect those outside China: Sourcify CEO.* Retrieved from https://www.cnbc.com/video/2020/01/28/coronavirus-supply-chain-disruption-will-affect-those-outside-china.html

Ozgur, C., Kumar, S., & Shen, Y. (2018). *The effect of supply chain disruption on average lateness and tardiness the affect of average tardiness on supply chain disruptions.*

Park, J., Shin, K. Chang, T.-W., & Park, J. (2010). An integrative framework for supplier relationship management. *Industrial Management & Data Systems, 110*(4), 495–515.

Paul, S. K., Moktadir, M. A., & Ahsan, K. (2021). Key supply chain strategies for the post-COVID-19 era: Implications for resilience and sustainability. *The International Journal of Logistics Management*, Vol. ahead-of-print, No. ahead-of-print. https://doi.org/10.1108/IJLM-04-2021-0238

Pérez, M. P., & Sánchez, A. M. (2001). Supplier relations and flexibility in the Spanish automotive industry. *Supply Chain Management: An International Journal, 6*(1), 29–38.

Pieh, C., Budimir, S., Delgadillo, J., Barkham, M., Fontaine, J. R., & Probst, T. (2021). Mental health during COVID-19 lockdown in the United Kingdom. *Psychosomatic Medicine, 83*(4), 328–337.

Queiroz, M. M., Telles, R., & Bonilla, S. H. (2019). Blockchain and supply chain management integration: A systematic review of the literature. *Supply Chain Management: An International Journal.*

Rajeev, A., Pati, R. K., Padhi, S. S., & Govindan, K. (2017). Evolution of sustainability in supply chain management: A literature review. *Journal of Cleaner Production, 162*, 299–314.

Rao, S., Gulley, A., Russell, M., & Patton, J. (2021). On the quest for supply chain transparency through blockchain: Lessons learned from two serialized data projects. *Journal of Business Logistics, 42*(1), 88–100.

Revilla, E., & Saenz, M. J. (2017). The impact of risk management on the frequency of supply chain disruptions a configurational approach. *International Journal of Operations & Production Management, 37*(5), 557–576.

Rico, R., Sanchez-Manzanares, M., Gil, F., & Gibson, C. B. (2008). Team coordination processes: A team knowledge-based approach. *Academy of Management Review, 33*(1), 163–185.

Ring, P. S., Van de Ven, A. H. (1994). Developmental processes of cooperative interorganizational relationships. *Academy of Management Review, 19*(1), 90–118.

Roggeveen, A. L., & Sethuraman, R. (2020). How the COVID-19 pandemic may change the world of retailing. *Journal of Retailing, 96*(2), 169.

Rossi, R., Socci, V., Talevi, D., Mensi, S., Niolu, C., Pacitti, F., ... Di Lorenzo, G. (2020). COVID-19 pandemic and lockdown measures impact on mental health among the general population in Italy. *Frontiers in Psychiatry*, 790.

Rossin, D. (2007). An exploratory analysis of information quality in supply chains: Efficient and responsive models. *Journal of Global Business Issues, 1*(2), 151–158.

Sajjad, A. (2021). The COVID-19 pandemic, social sustainability, and global supply chain resilience: A review. *Corporate Governance, 21*(2), 1–17.

Salem, T. J. (2018). *An investigation of key risks and risk management strategies in construction projects—Gaza Strip.* M.Sc. in Civil Engineering Thesis, The Islamic University-Gaza.

Sanjay Kumar. (7/2016). Advance warning of supply chain disruption: A behavioral experiment. In *11th Annual Behavioral Operations Management Conference, at University of Wisconsin-Madison, Madison* https://www.researchgate.net/publication/305771646_Advance_Warning_of_Supply_Chain_Disruption_A_Behavioral_Experiment

Sarkis, J., & Talluri, S. (2002). A model for strategic selection. *Journal of Supply Chain Management, 38*(1), 18–28.

Sawik, T. *Supply chain disruption management © 2020 using stochastic mixed integer programming.* Supply Chain Disruption Management. SpringerLink

Schmitt, T., Kumar, S., Stecke, K., Glover, F., & Ehlen, M. (4/2017). Mitigating disruptions in a multi-echelon supply chain using adaptive ordering. *Omega, 68*, 185–198.

SCRC SME. (2004, September 28). *Supply chain resource cooperative*. Retrieved from https://scm.ncsu.edu/scm-articles/article/supply-chain-disruptions

Section 3: 21st century manufacturing: Supplier alliances—Chrysler and MAGNA international. (1997). *International Journal of Physical Distribution & Logistics Management, 27*(2), 117–119.

Shahed, K. S., Azeem, A., Ali, S. M., et al. (2021). A supply chain disruption risk mitigation model to manage COVID-19 pandemic risk. *Environmental Science and Pollution Research*. https://doi.org/10.1007/s11356-020-12289-4

Shahzad, A., Wenyu, C., & Kumar, R. (2021, April). Blockchain based monitoring on trustless supply chain processes. In *2021 IEEE 6th International Conference on Cloud Computing and Big Data Analytics (ICCCBDA)* (pp. 216–221). IEEE.

Sharma, A., Adhikary, A., & Borah, S. B. (2020). Covid-19's impact on supply chain decisions: Strategic insights from NASDAQ 100 firms using Twitter data. *Journal of Business Research, 117*, 443–449.

Shihua, M., & Xubin, W. (2002). A method of confirming the weight of attributes for supplier evaluation. *Industrial Engineering Management, 6*, 5–8.

Shin, H., Collier, D. A., & Wilson, D. D. (2000). Supply management orientation and supplier/buyer performance. *Journal of Operations Management, 18*(3), 317–333.

Shuyong, K., & Rongqiu, C. (1998). The relationship of manufacturers and suppliers under JIT. *Journal of Management Engineering, 3*, 46–51.

Skipper, J. B., & Hanna, J. B. (2009, June 12). *International Journal of Physical Distribution & Logistics Management*. ISSN: 0960-0035.

Smith, K. T., Smith, M., & Wang, K. (2010). Does brand management of corporate reputation translate into higher market value? *Journal of Strategic Marketing, 18*(3), 201–221.

Srinivas, S. S., & Marathe, R. R. (2021). Moving towards "mobile warehouse": Last-mile logistics during COVID-19 and beyond. *Transportation Research Interdisciplinary Perspectives, 10*, 100339.

Standaert, W., Muylle, S., & Basu, A. (2021). How shall we meet? Understanding the importance of meeting mode capabilities for different meeting objectives. *Information & Management, 58*(1), 103393.

Sterman, J. (2000). *Business dynamics: Systems thinking and modeling for a complex world* (p. 564). Irwin/McGraw-Hill.

Stevenson, W. J. (2018). *Operations management*. McGraw-Hill/Irwin.

Swan, M. (2015). *Blockchain: Blueprint for a new economy*. O'Reilly Media, Inc.

Szulanski, F., & Viñegla, A. L. (2003). XSC o Balanced Scorecard basado en laexperiencia y las vivencias de la red de valor. *Estrategia financiera, 194*, 38–44.

Takeishi, A. (2001). Bridging inter-and intra-firm boundaries: Management of supplier involvement in automobile product development. *Strategic Management Journal, 21*(5), 403–433.

Tang, C.-H., Chin, C.-Y., & Lee, Y.-H. (2021). Coronavirus disease outbreak and supply chain disruption: Evidence from Taiwanese firms in China. *Research in International Business and Finance, 56*, 101355. ISSN 0275-5319.

Tennenhouse, E. (2017). It's a no-brainer. *New Scientist, 235*(3134), 32–35.

Tomlin, B. (2006). On the value of mitigation and contingency strategies for managing supply chain disruption risks. *Management Science, 52*(5), 639–657.

Tracey, M., & Tan, C. L. (2001). Empirical analysis of supplier selection and involvement, customer satisfaction, and firm performance. *Supply Chain Management: An International Journal, 6*(4), 174–188.

Ulaga, W., & Eggert, A. (2006). Value-based differentiation in business relationships: Gaining and sustaining key supplier status. *Journal of Marketing, 70*(1), 119–136.

Ustun, O., & Demirtas, E. A. (2008). Multi-period lot-sizing with supplier selection using achievement scalarizing functions. *Computers, and Industrial Engineering, 54*(4), 918–931.

Vekstein, D. (1998). Managing knowledge and corporate performance: An empirical analysis of the world automobile industry. *Omega International Journal of Management Science, 26*(5), 551–568.

Veselovská, L. (2020). Supply chain disruptions in the context of early stages of the global COVID-19 outbreak. *Problems and Perspectives in Management, 18*(2), 490–500. https://doi.org/10.21511/ppm.18(2).2020.40

Walling, C. (2017, August 28). *5 strategies to protect against supply chain disruptions.* Retrieved from https://www.industrystarsolutions.com/blog/category/supply-chain-disruption/

Weber, C., Current, J. R., & Benton, W. C. (1991). Vendor selection criteria and methods. *European Journal of Operational Research, 50*(1), 2–19.

Wee, H. M., Yang, W. H., Chou, C. W., & Padilan, M. V. (2012). Renewable energy supply chains, performance, application barriers, and strategies for further development. *Renewable and Sustainable Energy Reviews, 16*(8), 5451–5465.

WGA Consulting. (2017, September 18). *Supply chain disruption—Reducing the impact and cost.* Retrieved from https://www.wgaconsulting.com/new/2016/02/24/supply-chain-disruption-reducing-impact-cost-supply-chain-disruption

Wheatley, M., & Frieze, D. (2011). *Walkout, walk on.* Berrett-Koehler Publishers.

Whyte, D. (2002). *Crossing the unknown sea.* Riverhead Books.

Wiengarten, F., Humphreys, P., Cao, G., Fynes, B., and McKittrick, A. (2010). Collaborative supply chain practices and performance: Exploring the key role of information quality. *Supply Chain Management: An International Journal, 15*(6), 463–473.

Wildgoose, N., Brennan, P., & Thompson, S. (2012). Understanding your supply chain to reduce the risk of supply chain disruption. *Journal of Business Continuity & Emergency Planning, 6*(1), 55–67.

Wu, T., Blackhurst, J., & O'grady, P. (2007). Methodology for supply chain disruption analysis. *International Journal of Production Research, 45*(7), 1665–1682. https://doi.org/10.1080/00207540500362138

Xia, W., & Wu, Z. (2007). Supplier selection with multiple criteria in volume discount environments. *Omega, 35*(5), 494–504.

Yun, M. (1999). Subcontracting relations in the Korean automotive industry: Risk sharing and technological capability. *International Journal of Industrial Organization, 17*(1), 81–108.

Zolfani, S. H., et al. (2012). A hybrid MCDM model encompassing AHP and COPRAS-G methods for selecting company suppliers in Iran. *Technological and Economic Development of Economy, 18*(3), 529–536.

Theme 2: Understanding Supply Chain Dynamics—Modeling-Based Empirical Solutions

Here, a total of 332 studies covering empirical studies, utilizing various modeling approaches especially system dynamics, relevant to various topics of this theme are included. Key topics addressed in these studies include Supply chain management practices, Organizational factors, Internet of things and supply chain management, *Supply chain management best practice*, Supply Chain Integrity, Food supply chains and Covid-19, Sustainable supply chain management, Global supply chains, Supply chains during Covid-19 pandemic, Food supply chain resilience, Green supply chain management, Sustainable Innovation, Supply chain management practices in state-owned enterprises, Logistics and ecological supply chain performance, Talent management, Supply chain disruptions and challenges, Environmental purchasing and supplier management, Sustainability strategies, Managing supply chain risk, International supplier network, Collaboration and logistics efficiency, Pharmaceutical industry supply planning, Supply chain integration, Supplier-retailer collaboration, Supply chain uncertainty and risk, The triple bottom line, Reverse supply chain investments, Sustainable supply chain management, Agile supply-chain management, COVID-19, supply chain management alignment.

References for Theme 2

Abernethy, M. A., & Chua, W. F. (1996). A field study of control redesign: The impact of institutional processes on strategic choice. *Contemporary Accounting Research, 13*(2), 569–606.

Acs, Z., & Armington, C. (2004). Employment growth and entrepreneurial activity in cities. *Regional Studies, 38*, 911–927.

Adyel, T. M. (2020). Accumulation of plastic waste during COVID-19. *Science, 369*(6509), 1314–1315.

Ahlqvist, V., Norrman, A., & Jahre, M. (2021). Supply chain risk governance: Towards a conceptual multi-level framework. *Operations and Supply Chain Management, 13*, 382–395.

Ahmad, N., & Seymour, R. (2008). *Defining entrepreneurial activity: Definitions supporting frameworks for data collection.*

Ahuja, V., & Sen, A. (2007). *Scope and space for small scale poultry production in developing countries.* Indian Institute of Management.

Akaike, H. (1987). Factor analysis and AIC. *Psychometrika, 52*(3), 317–332.

Akter, S., & Farrington, J. (2007). *Sustainability of an innovation for poverty alleviation: The case of Bangladesh Poultry Model.*

Akter, S., & Farrington, J. (2008, July 8). Poverty transition through targeted programme: The case of Bangladesh Poultry Model. In *82nd Annual Conference of the Agricultural Economics Society. Royal Agricultural College.* UK. http://ageconsearch.umn.edu/bitstream/36765/2/Akter_farrington.pdf

Akter, S., & Farrington, J. (2009). What makes exit from poverty: Investigation of smallholder women livestock farmers in Bangladesh. 2009 Conference, August 16–22, 2009, Beijing, China. International Association of Agricultural Economists.

Alam, G. M., Hoque, K. E., Khalifa, M. T. B., Siraj, S. B., & Ghani, M. F. B. A. (2009). The role of agriculture education and training on agriculture economics and national development of Bangladesh. *African Journal of Agricultural Research, 4*, 1334–1350.

Al-Doori, J. A. (2019). The impact of supply chain collaboration on performance in the automotive industry: Empirical evidence. *Journal of Industrial Engineering and Management, 12*(2), 241–253.

Ali, S. A., Baloch, M., Ahmed, N., Ali, A. A., & Iqbal, A. (2020). The outbreak of coronavirus disease 2019 (COVID-19)—An emerging global health threat. *Journal of Infection and Public Health, 13*(4), 644–646.

Alkahtani, M., Omair, M., Khalid, Q. S., Hussain, G., Ahmad, I., & Pruncu, C. (2021). A COVID-19 supply chain management strategy based on variable production under uncertain environment conditions. *International Journal of Environmental Research and Public Health, 18*, 1662.

Al-Shboul, M. A. R., Barber, K. D., Garza-Reyes, J. A., Kumar, V., & Abdi, M. R. (2017). The effect of supply chain management practices on supply chain and manufacturing firms' performance. *Journal of Manufacturing Technology Management, 28* (5), 577–609.

Anderson JC, Gerbing DW (1988) Structural equation modeling in practice: a review and recommended two-step approach. Psychological Bulletin 103(3):411–423.

Anparasan, A. A., & Lejeune, M. A. (2018). Data laboratory for supply chain response models during epidemic outbreaks. *Annals of Operations Research, 270*, 53–64.

Antony, J., Psomas, E., Garza-Reyes, J. A., & Hines, P. (2021) Practical implications and future research agenda of lean manufacturing: A systematic literature review. *Production Planning & Control, 32*, 889–925.

Asaduzzaman, M. (2000). *Livestock sector, economic development and poverty alleviation in Bangladesh: Changing rural economy of Bangladesh* (pp. 42–53). Bangladesh Economic Association.

Åstebro, T., & Thompson, P. (2011). Entrepreneurs, jacks of all trades or Hobos? *Research Policy, 40*, 637–649.

Azadegan, A., & Dooley, K. (2021). A typology of supply network resilience strategies: Complex collaborations in a complex world. *Journal of Supply Chain Management, 57*(1), 17–26.

Azevedo, A., & Sousa, J. P. (2000). Agile supply-chain management: Challenges, requirements and solutions. *IFAC P, 33*, 917–922.

Bag, S., Dhamija, P., Luthra, S., & Huisingh, D. (2021). How big data analytics can help manufacturing companies strengthen supply chain resilience in the context of the COVID-19 pandemic. *The International Journal of Logistics Management*. https://doi.org/10.1108/IJLM-02-2021-0095

Bakos, L., & Dumitras, D. D. (2021). Decentralized enterprise risk management issues under rapidly changing environments. *Risks, 9*, 165.

Balci, O. (1994). Validation, verification, and testing techniques throughout the life cycle of a simulation study. *Annals of Operations Research, 53*, 121–173.

Banaitiene, N., & Banaitis, A. (2012). Risk management in construction projects. In N. Banaitiene (Ed.), *Risk management: Current issues* (pp. 429–448). InTech Open.

Barclay, I. (2002). Organizational factors for success in new product development. *IEEE Proceedings—Science, Measurement and Technology, 149*(2), 105–112.

Barlas, Y. (1989). Multiple tests for validation of system dynamics type of simulation models. *European Journal of Operational Research, 42*, 59–87.

Barlas, Y. (1996). Formal aspects of model validity and validation in system dynamics. *System Dynamics Review, 12*, 183–210.

Barratt, M. (2004). Understanding the meaning of collaboration in the supply chain. *Supply Chain Management: An International Journal, 9*, 30–42.

Barrera-Roldán, A., & Saldivar-Valdés, A. (2002). Proposal and application of a Sustainable Development Index. *Ecological Indicators, 2*, 251–256.

Bass, F. M. (1969). A new product growth model for consumer durables. *Management Science, 15*, 215–227.

BBS. (2010). *Rural poverty portal* [Online]. Dhaka. http://www.ruralpovertyportal.org/web/guest/country/statistics/tags/bangladesh. Accessed 29 Mar 2012.

Belhadi, A., Kamble, S., Jabbour, C. J. C., Gunasekaran, A., Ndubisi, N. O., & Venkatesh, M. (2021). Manufacturing and service supply chain resilience to the COVID-19 outbreak: Lessons learned from the automobile and airline industries. *Technological Forecasting and Social Change, 163*, 120447.

Belisle, P. T. (2011). Sustainability/climate change. *Strategic Planning for Energy and the Environment, 30*, 71–78.

Ben-Daya, M., Hassini, E., & Bahroun, Z. (2019). Internet of things and supply chain management: A literature review. *International Journal of Production Research, 57*(15–16), 4719–4742.

Bhattacharay, S., Smark, C., & Mir, M. (2021). Covid-19: Social, financial and economic implications. *AABFJ, 15*(1), 1–4.

Birkel, H. S., & Hartmann, E. (2019) Impact of IoT challenges and risks for SCM. *Supply Chain Management, 24*, 39–61.

Bjerke, B. (2007). *Understanding entrepreneurship*. Elgar.

Blanchard, D. (2021). *Supply chain management best practices* (3rd ed.). Wiley.

Blanchard, O., & Portugal, P. (2001). What hides behind an unemployment rate: Comparing Portuguese and U.S. labor markets. *American Economic Review, 91*(1), 187–207.

Bocanet, A., Alpenidze, O., & Badran, O. (2021). Business analysis in post-pandemic era. *Academy of Strategic Management Journal, 4*, 1–9.

Boiral, O. (2006). La certification ISO 14001: une perspective néo institutionnelle. *Management International, 10*(3), 67–79.

Boström, M., Jönsson, A. M., Lockie, S., Mol, A. P., & Oosterveer, P. (2015). Sustainable and responsible supply chain governance: Challenges and opportunities. *Journal of Cleaner Production, 107*, 1–7.

Bromley, R., & Wilson, T. D. (2018). The urban informal economy revisited. *Latin American Perspectives, 45*(1, 128), 4–23.

Butler, C. D. (2000). Global Change Hum. *Health, 1*, 156.

Bynum, C., Sze, C., Kearns, D., Polovick, B., & Simon, K. (2018). An examination of a voluntary policy model to effect behavioral change and influence interactions and decision making in the freight sector. *Transportation Research Part D, 61*, 19–32.

Calatayud, A., Mangan, J., & Christopher, M. (2019). The self-thinking supply chain. *Supply Chain Management, 24*, 22–38.

Calina, D., Docea, A. O., Petrakis, D., Egorov, A. M., Ishmukhametov, A. A., Gabibov, A. G., Shtilman, M. L., Kostoff, R., Caralho, F., Vinceti, M., Spandidios, D. A., & Tsatsakis, A. (2020). Towards effective covid-19 vaccines: Updates perspectives and challenges (review). *International Journal of Molecular Medicine, 46*, 3–16.

Casey, T. R., & Töyli, J. (2012). Dynamics of two-sided platform success and failure: An analysis of public wireless local area access. *Technovation, 32*, 703–716.

Castillo, V. E., Mollenkopf, D. A., Bell, J. E., & Bozdogan, H. (2018). Supply chain integrity: A key to sustainable supply chain management. *Journal of Business Logistics, 39*(1), 38–56.

Castka, P., & Balzarova, M. (2008). ISO 26000 and supply chains on the diffusion of the social responsibility standard. *International Journal of Production Economics, 111*, 274–286.

Chaudhuri, A., Boer, H., & Taran, Y. (2018). Supply chain integration, risk management, and manufacturing flexibility. *Journal of Operations & Production Management*.

Chen, C.-N., Tzeng, L.-C., Ou, W.-M., & Chang, K.-T. 2007. The relationship among social capital, entrepreneurial orientation, organizational resources and entrepreneurial performance for new ventures. *Contemporary Management Research*, 3.

Chen, L., Zhao, X., Tang, O., Price, L., Zhang, S., & Zhu, W. (2017). Supply chain collaboration for sustainability: A literature review and future research agenda. *Journal of Production Economics, 194*, 73–87.

Cheong, J., Kwak, D. W., & Tang, K. K. (2018). The trade effects of tariffs and non-tariff changes of preferential trade agreements. *Economic Modelling, 70*, 370–382.

Chopra, S. (2019). *Supply chain management: Strategy, planning, and operation* (7th ed.). Pearson.

Christopher, M. (2000). The agile supply chain: Competing in volatile markets. *Industrial Marketing Management, 29*, 37–44.

Churchman, C. W. (1968). *The systems approach*. Dell Publishing Co., Inc.

Clapp, J., & Moseley, W. G. (2020). The food crisis is different: Covid-19 and the fragility of the neoliberal food security order. *Journal of Peasant Studies, 47*(7), 1393–1417.

Cobb, C., Schuster, D., Beloff, B., & Tanzil, D. (2009). The AIChE sustainability index: The factors in detail. *Chemical Engineering Progress, 105*, 60.

Cohen, L., Manion, L., & Morrison, K. (2011). *Research methods in education*. Routledge.

Collier, Z. A., & Sarkis, J. (2021). The zero trust supply chain: Managing supply chain risk in the absence of trust. *Journal of Production Research, 59*(11), 3430–3445.

Cooper, C. M., Lambert, M. D., & Pagh, D. J. (1997). Supply chain management: More than a new name for logistics. *International Journal of Logistics Management, 8*, 1–9.

Coulthard, S., Johnson, D., & Mcgregor, J. A. (2011). Poverty, sustainability and human wellbeing: A social wellbeing approach to the global fisheries crisis. *Global Environmental Change, 21*, 453–463.

Council, S.-C. (1999). *Supply-chain council*. The Council.

Cox, J. F., Blackstone, J. H., & Spencer, M. S. (1995). *APICS dictionary* (8th ed.). American Production and Inventory Control Society.

Craig, C. R., & Dale, R. S. (2008). A framework of sustainable supply chain management: Moving toward new theory. *International Journal of Physical Distribution & Logistics Management, 38*, 360–387.

Cranfield, J. A. L. (2020). Framing consumer food demand responses in a viral pandemic. *Canadian Journal of Agricultural Economics, 68*, 151–156.

Cresswell, J. W. (2003). *Research design: Quantitative, qualitative & mixed method approaches, United States*. Sage.

Dadhich, P., Genovese, A., Kumar, N., & Acquaye, A. (2015). Developing sustainable supply chain in the UK construction industry: A case study. *International Journal of Production Economics, 164*, 271–284.

Daily, G. C., & Ehrlich, P. R. (1992). Population, sustainability, and Earth's carrying capacity. *BioScience, 42*, 761–771.

Danese, P., Romano, P., & Formentini, M. (2013). The impact of supply chain integration on responsiveness: The moderating effect of using an international supplier network. *Transportation Research Part E: Logistics and Transportation Review, 49*(1), 125–140.

Davis, J. L., Bell, R. G., Payne, G. T., & Kreiser, P. M. (2010). Entrepreneurial orientation and firm performance: The moderating role of managerial power. *American Journal of Business, 25*, 41–54.

Davis, K. (1990). Population and resources: Fact and interpretation. *Population and Development Review, 16*, 1–21.

Davydenko, I., Ehrler, V., de Ree, D., Lewis, A., & Tavasszy, L. (2014). Towards a global CO_2 calculation standard for supply chains: Suggestions for methodological improvements. *Transportation Research Part D: Transport and Environment, 32*, 362–372.

De Campos Silva, A. R. (2020). Health risks for poultry workers in Brazil in the covid-19 pandemic. *Bulletin of Latin American Research, 39*(S1), 88–91.

De La Fuente, M., Ros, L., & Ortiz, A. (2010). Enterprise modelling methodology for forward and reverse supply chain flows integration. *Computers in Industry, 61*, 702–710.

Deconinck, K., Avery, E., & Jackson, L. A. (2021). Food supply chains and covid-19: Impacts and policy lessons. *EuroChoices, 19*(3), 34–39.

Dees, J. G. (1998). *The meaning of social entrepreneurship*. Comments and suggestions contributed from the Social Entrepreneurship Funders Working Group, 6pp.

Dhama, K., Khan, S., Tiwari, R., Sircar, S., Bhat, S., Malik, Y. S., Singh, K. P., Chaicumpa, W., Bonilla-Aldana, D. K., & Rodriguez-Morales, A. J. (2020). Coronavirus disease 2019—COVID-19. *Clinical Microbiology Reviews, 33*, e00028–20.

Di Francesco, R. M., Meena, P., & Tibrewala, R. (2021). Buyback and risk-sharing contracts to mitigate the supply and demand disruption risks. *European Journal of Industrial Engineering, 15*, 550–581.

Diaz, E., Mamelund, S., Eid, J., Aasen, H. S., Kaarboe, O. M., Brokstad, R. J. C., Gloppen, S., Beyer, A., & Kumar, B. N. (2021). Learning from the covid-19 pandemic among migrants: An innovative, system-level, interdisciplinary approach is needed to improve public health. *Scandinavian Journal of Public Health, 49*, 804–808.

Dictionary. (2021). *Dictionary*. Retrieved October 12, 2021, from https://www.dictionary.com/browse/trash

DiMaggio, P. J., & Powell, W. (1983). The iron cage revisited institutional isomorphism and collective rationality in organizational fields. *American Sociological Review, 48*, 147–160.

Dolberg, F. (2004). *Review of household Poultry production as a tool in poverty reduction with focus on Bangladesh and India, India, National Dairy Development Board.*

Downs, A., & Acevedo, R. (2019). *How our trash impacts the environment*. Retrieved October 12, 2021, from https://www.earthday.org/how-our-trash-impacts-the-environment/

Durmaz, A., Demir, H., & Sezen, B. (2021). The role of negative entropy within supply chain sustainability. *Sustainable Production and Consumption, 28*, 218–230.

Dyer, J. H., & Chu, W. (2003). The role of trustworthiness in reducing transaction costs and improving performance: Empirical evidence from the United States, Japan, and Korea. *Organization Science, 14*, 57–68.

Ecube Labs. (2016). *Overflowing garbage bins: 5 impacts on health and environment, and how to prevent*. Retrieved September 4, 2021, from https://www.ecubelabs.com/overflowing-garbage-bins-5-impacts-on-health-and-environment-and-how-to-prevent/

Eisenmann, T., Parker, G., & Van Alstyne, M. (2011). Platform envelopment. *Strategic Management Journal, 32*(12), 1270–1285. https://doi.org/10.1002/smj.935

El Baz, J., & Ruel, S. (2021). Can supply chain risk management practices mitigate the disruption impacts on supply chains' resilience and robustness? Evidence from an empirical survey in a COVID-19 outbreak era. *International Journal of Production Economics, 233*, 107972.

Elkington, J. (1994). Towards the sustainable corporation: Win-win-win business strategies for sustainable development. *California Management Review, 36*, 90–100.

Elkington, J. (2004). Enter the triple bottom line. *The Triple Bottom Line: Does It All Add Up*, 1–16.

Ellram, L., & Cooper, M. (1993). Characteristics of supply chain management and the implications for purchasing and logistics strategy. *International Journal of Logistics Management, 4*, 1–10.

Emrouznejad, A., & Marra, M. (2017). The state of the art development of AHP (1979– 2017): A literature review with a social network analysis. *International Journal of Production Research, 55*, 6653–6675.

EPA. (2020). *EPA United States Environmental Protection Agency*. Retrieved October 12, 2021, from https://www.epa.gov/rcra/medical-waste

Erol, I., & Velioglu, M. N. (2019). An investigation into sustainable supply chain management practices in a developing country. *International Journal of Business and eGovernment Studies, 11*(2), 104–118.

Esper, T. L. (2021). Supply chain management amid the coronavirus pandemic. *Journal of Public Policy & Marketing, 40*(1), 101–102.

Fairclough, N. (2003). *Analysing discourse: Textual analysis for social research*. Routledge.

Farooq, M. U., Hussain, A., Masood, T., & Habib, M. S. (2021). Supply chain operations management in pandemics: A state-of-the-art review inspired by COVID-19. *Sustainability, 13*, 2504.

Fatka, J. (2021a). USDA inspector to examine meat processing COVID cases. *Western Farm Press-Penton Business Media, 1*(1), 1–3.

Fatka, J. (2021b). *Higher food prices could result if meat plants are required to increases space requirements between workers*.

Febianto, R. (2021). *Jumlah sampah melonjak selama pandemi COVID-19, ini beberapa penyebabnya*. Retrieved September 4, 2021, from https://www.liputan6.com/health/read/4483554/jumlah-sampah-melonjak-selama-pandemi-covid-19-ini-beberapa-penyebabnya

Fisher, M. L. (1997, May–April). What is the right supply chain for your product? Harvard Business Review, 1–11.

Flynn, B. B., Huo, B., & Zhao, X. (2010). The impact of supply chain integration on performance: A contingency and configuration approach. *Journal of Operations Management, 28*(1), 58–71.

Flynn, B., Cantor, D., Pagell, M., Dooley, K. J., & Azadegan, A. (2021). From the editors: Introducing to managing supply chains beyond covid 19—Preparing for the next global mega disruption. *Journal of Supply Chain Management,* 57(1), 3–6.

Flynn, B., Cantor, D., Pagell, M., Dooley, K. J., & Azadegan, A. (2021). From the editors: Introduction to managing supply chains beyond COVID-19—Preparing for the next global mega-disruption. *Journal of Supply Chain Management, 57*, 3–6.

Folia, R. (2019). *Eva Bachtiar, Pahlawan Sampah Makanan di Surabaya*. Retrieved September 3, 2021, from https://www.idntimes.com/life/women/rosa-folia/eva-bachtiar-pahlawan-sampah-makanan-di-surabaya-1/2

Fonseca, L. M., & Azevedo, A. L. (2020). Covid-19: Outcomes for global supply chains. *Management & Marketing Challenges for the Knowledge Society, 15*(Special issue).

Ford, A. (1999). *Modeling the environment: An introduction to system dynamics models of environmental systems*. Island Press.

Forrester, J. W. (1961). *Industrial dynamics*. MIT Press, currently available from Pegasus Communications.

Forrester, J. W. (1999). *System dynamics: The foundation under system thinking*. Sloan School of Management MIT.

Forrester, J. W., & Senge, P. M. (1980). Tests for building confidence in system dynamics models. *TIMS Studies in Management Sciences, 14*, 209–228.

Freytag, A., & Thurik, R. (2010). *Entrepreneurship and its determinants in a cross-country setting*. Springer.

Friday, D., Savage, D. A., Melnyk, S. A., Harrison, N., Ryan, S., & Wechtler, H. (2021). A collaborative approach to maintaining optimal inventory and mitigating stockout risks during a pandemic: Capabilities for enabling health-care supply chain resilience. *European Journal of Industrial Engineering, 11*, 248–271.

Gawade, D. (2021). Agile supply chain in manufacturing and service industry: Bibliometric and content analysis. *Vision*. https://doi.org/10.1177/09722629211002000

Gerbing, D. W., & Anderson, J. C. (1988). An updated paradigm for scale development incorporating unidimensionality and its assessment. *Journal of Marketing Research, 25*(2), 186–192.

Global Data. (2021). *Indonesia's e-commerce market continues to surge amid the COVID-19 pandemic*. https://www.globaldata.com/indonesias-e-commerce-market-continues-surge-amid-covid-19-pandemic-says-globaldata/

Göçer, F. (2021). A novel interval value extension of picture fuzzy sets into group decision making: An approach to support supply chain sustainability in catastrophic disruptions. *IEEE Access, 9*, 117080–117096.

Gölgeci, I., & Kuivalainen, O. (2020). Does social capital matter for supply chain resilience? The role of absorptive capacity and marketing–supply chain management alignment. *Industrial Marketing Management, 84*, 63–74.

Goodland, R., & Daly, H. (1996). Environmental sustainability: Universal and non-negotiable. *Ecological Applications*, 1002–1017.

Gralinski, L. E., & Menachery, V. D. (2020). Return of the coronavirus: 2019-nCov. *Viruses, 12*(135), 1–8.

Gri, G. R. I. (2009). *What is sustainability reporting*. Retrieved August 14, 2009.

Grönroos, C. (1997). Value-driven relational marketing: From products to resources and competencies. *Journal of Marketing Management, 13*, 407–419.

Gunasekaran, A., Lai, K., & Cheng, T. C. E. (2008). Responsive supply chain: A competitive strategy in a networked economy. *The International Journal of Management Science, 36*, 549–564.

Guo, Y., Cao, Q., Hong, Z., Tan, Y., Chen, S., Jin, H., Tan, K., Wang, D., & Yan, Y. (2020). The origin, transmission and clinical therapies on coronavirus disease 2019 (COVID-19) outbreak—An update on the status. *Military Medical Research, 7*(11), 1–10.

Ha, B. C., Park, Y. K., & Cho, S. (2011). Suppliers' affective trust and trust incompetency in buyers: Its effect on collaboration and logistics efficiency. *International Journal of Operations & Production Management, 31*(1–2), 56–77. https://doi.org/10.1108/01443571111098744

Hafez, H. M., & Attia, Y. A. (2020). Challenges to the poultry industry: Current perspectives and strategic future after the covid-19 outbreak. *Frontiers in Veterinary Science, 7*(516), 1–16.

Hair, J. F. J., Black, W. C., Babin, B. J., Anderson, R. E., Tatham, R. L. (2006). *Multivariate data analysis*. Prentice-Hall.

Hajiagha, S. H. R., Mahdiraji, H. A., Behnam, M., Nekoughadirli, B., & Joshi, R. (2021). A scenario-based robust time–cost tradeoff model to handle the effect of COVID-19 on supply chains project management. *Operations Management Research*. https://doi.org/10.1007/s12063-021-00195-y

Hall, J. K., Daneke, G. A., & Lenox, M. J. (2010). Sustainable development and entrepreneurship: Past contributions and future directions. *Journal of Business Venturing, 25*, 439–448.

Hallavo, V. (2015). Superior performance through supply chain fit: A synthesis. *Supply Chain Management: An International Journal, 20*(1), 71–82.

Hancock, G. (1992). *Lords of poverty: The power, prestige, and corruption of the international aid business*. Atlantic Monthly Press.

Harapko, S. (2021). How COVID-19 impacted supply chains and what comes next. https://www.ey.com/en_id/supply-chain/how-covid-19-impacted-supply-chains-and-what-comes-next

Harvey, C., & Lewis, S. R. (1991). Policy choice and development performance in Botswana. *African Studies Review, 34*(3), 137–139.

Hayakawa, K., & Mukunoki, H. (2021). The impact of covid-19 on international trade: Evidence from the first shock. *Journal of the Japanese and International Economies, 60*, 101–135.

Heck, R. K., & Stafford, K. (2001). The vital institution of family business: Economic benefits hidden in plain sight. *Destroying Myths and Creating Value in Family Business*, 9–17.

Hidranto, F. (2021). *Bisnis e-commerce semakin gurih*. https://www.indonesia.go.id/kategori/indonesia-dalam-angka/2534/bisnis-e-commerce-semakin-gurih

Hobbs, J. E. (2020). Food supply chains during Covid-19 pandemic. *Canadian Journal of Agricultural Economics, 68*, 171–176.

Hobbs, J. E. (2021). Food supply chain resilience and the covid-19 pandemic: What have we learned? *Canadian Journal of Agricultural Economics, 69*, 189–196.

Holweg M (2005) An investigation into supplier responsiveness: Empirical evidence from the automotive industry. Int J Logist Manag 16:96–119

Homer, J., & Oliva, R. (2001). Maps and models in system dynamics: a response to Coyle. *System Dynamics Review, 17*, 347–355.

Honig, B. (1998). What determines success? Examining the human, financial, and social capital of Jamaican microentrepreneurs. *Journal of Business Venturing, 13*, 371–394.

Hope, K. R. (1995). Managing the public sector in Botswana. *International Journal of Public Sector Management, 8*(6), 51–62.

https://preventepidemics.org/covid19/science/insights/update-on-covid-19-in-africa/

Hwang, C. L., & Yoon, K. (1981). *Multiple attribute decision making: Methods and applications: A state-of-the-art survey*. Springer.

ILO. (2020, May). *International labour organization (ILO) covid-19 monitor* (4th ed.). https://www.ilo.org/wcmsp5/groups/public/---dgreports/---dcomm/documents/briefingnote/wcms_745963.pdf

ILO. (2021). *ILO monitor: Covid-19 and the world of work* (8th ed.). International Labour Organization. https://www.ilo.org/wcmsp5/groups/public/---dgreports/---dcomm/documents/briefingnote/wcms_824092.pdf

Insider, L. (2021). *Household waste definition*. Retrieved October 9, 2021, from https://www.lawinsider.com/dictionary/household-waste

Ishizaka, A., & Nemery, P. (2013). *Multi-criteria decision analysis: Methods and software*. Wiley.

Ismail, H. S., & Sharifi, H. (2006). A balanced approach to building agile supply chains. *International Journal of Physical Distribution & Logistics Management, 36*, 431–444.

Ivanov, D. (2020). Predicting the impacts of epidemic outbreaks on global supply chains: A simulation-based analysis on the coronavirus outbreak (COVID-19/SARS-CoV-2) case. *Transportation Research Part E: Logistics and Transportation Review, 136*, 101922.

Ivanov, D., & Dolgui, A. (2019). New disruption risk management perspectives in supply chains: Digital twins, the ripple effect, and resileanness. *IFAC Papersonline, 52*, 337–342.

Ivanov, D., & Dolgui, A. (2020). OR-methods for coping with the ripple effect in supply chains during COVID-19 pandemic: Managerial insights and research implications. *International Journal of Production Economics, 232*, 107921.

Jabbarzadeh, A., Fahimnia, B., & Sabouhi, F. (2020). Resilient and sustainable supply chain design: Sustainability analysis under disruption risks. *International Journal of Production Research, 56*, 5945–5968

Jabbour, A. B. L. S., Song, M., & Godinho Filho, M. (2020). Sustainability implications for operations management: Building the bridge through exemplar case studies. *Production Planning & Control, 31*, 841–844.

Jain, V., Wadhwa, S., & Deshmukh, S. (2009). Select supplier-related issues in modelling a dynamic supply chain: Potential, challenges and direction for future research. *International Journal of Production Research, 47*, 3013–3039.

Jennings, P. D., & Zandbergen, P. A. (1995). Ecologically sustainable organizations: An institutional approach. *The Academy of Management Review, 20*, 1015–1052.

Johnson, R. B., & Onwuegbuzie, A. J. (2004). Mixed methods research: A research paradigm whose time has come. *Educational Researcher, 33*, 14–26.

Jørgensen, H. D. (2004). *Interactive process models*. Norwegian University of Science and Technology.

Juan Ding, M., Jie, F. A., Parton, K., & Matanda, M. J. (2014). Relationships between quality of information sharing and supply chain food quality in the Australian beef processing industry. *The International Journal of Logistics Management, 25*(1), 85–108.

Karthik, K. A., & Sinha, M. (2021). The impact of physical distancing on the sharing economy. *Australasian Accounting, Business and Finance Journal, 15*(1), 22–36.

Kementerian Kesehatan RI. (2015). *Rencana Strategis Kementerian Kesehatan (2015–2019)*. Kementerian Kesehatan RI.

Kersan-Skabic, I. (2021). The covid-19 pandemic and the internationalization of production: A review of the literature. *Development Policy Review, 40*(2), 1–15.

Khan, S. A., Chaabane, A., & Dweiri, F. T. (2018). Multi-criteria decision-making methods application in supply chain management: A systematic literature review. In V. Salomon (Ed.), *Multi-criteria methods and techniques applied to supply chain management* (pp. 3–31). InTech Open.

Kiely, D. (2004). The state of pharmaceutical industry supply planning and demand forecasting. *The Journal of Business Forecasting, 23*(3), 20–22.

Kim, Y., Choi, T. Y., Yan, T., & Dooley, K. (2011). Structural investigation of supply networks: A social network analysis approach. *Journal of Operations Management, 29*(3), 194–211.

Kline, R. B. (2015). *Principles and practice of structural equation modeling*. New York: Guilford Publications.

Knoepfel, I. (2001). Dow Jones sustainability group index: A global benchmark for corporate sustainability. *Corporate Environmental Strategy, 8*, 6–15.

Kocabasoglu, C., Prahinski, C., & Klassen, R. D. (2007). Linking forward and reverse supply chain investments: The role of business uncertainty. *Journal of Operations Management, 25*, 1141–1160.

Krantz, L. (2001). *The sustainable livelihood approach to poverty reduction: An introduction*. SIDA.

Kreiser, P. M., & Davis, J. (2010). Entrepreneurial orientation and firm performance: The unique impact of innovativeness, proactiveness, and risk-taking. *Journal of Small Business & Entrepreneurship, 23*, 39–51.

Kumar, A., & Kushwaha, G. (2018). Supply chain management practices and operational performance of fair price shops in India: An empirical study. *Scientific Journal of Logistics, 14*(1), 85–99.

Kumar, A., Mangla, S. K., Kumar, P., & Song, M. (2021). Mitigate risks in perishable food supply chains: Learning from COVID-19. *Technological Forecasting and Social Change, 166*, 120643.

Kunc, M. H., & Morecroft, J. D. (2010). Managerial decision making and firm performance under a resource-based paradigm. *Strategic Management Journal, 31*(11), 1164–1182.

Lambert, D. M., & Cooper, M. C. (2000). Issues in supply chain management. *Industrial Marketing Management, 29*(1), 65–83.

Lancet, T. (2020). The plight of essential workers during the covid-19 pandemic. *Lancet, 395*, 1587.

Larraâin, J. (1979). *The concept of ideology*. University of Georgia Press.
Larue, B. (2021). Covid-19 and labor issues: An assessment. *Canadian Journal of Agricultural Economics, 69*(2), 269–279.
Lazear, E. (2003). *Entrepreneurship, IZA Disc. Paper*.
Lee, C., & Ha, B. C. (2020). The impact of interactional justice and supply-chain collaboration on sustainable SCM performance: The case of multinational pharmaceutical firms. *The Journal of Asian Finance, Economics and Business, 7*(2), 237–247.
Lee, H. L. (2004). The Triple-A supply chain: The 21st-century supply chain. *Harvard Business Review*.
Leite, H., Lindsay, C., & Kumar, M. (2020). COVID-19 outbreak: Implications on health-care operations. *The TQM Journal, 33*, 247–256.
Li, J., Wang, Y., Feng, G., Wang, S., & Song, Y. (2021). Supply chain finance review: Current situation and future trend. *Systems Engineering—Theory & Practice, 40*, 1977–1995
Lin, C. H., Chiu, H., & Chu, P. Y. (2005). Agility index in the supply chain. *International Journal of Production Economics, 100*, 285–299.
Loannou, I., & Serafeim, G. (2017). *The consequences of mandatory corporate sustainability reporting*. Harvard Business School Research Working Paper, n°11–100.
Lombardi Netto, A., Salomon, V. A. P., Ortiz-Barrios, M. A. (2021). Multi-criteria analysis of green bonds: Hybrid multi-method applications. *Sustainability, 13*, 10512.
Lummus, R. R., & Alber, K. L. (1997). *Supply chain management: Balancing the supply chain with customer demand*. APICS Educational and Research Foundation, Inc.
Lummus, R. R., & Vokurka, R. J. (1999). Defining supply chain management: A historical perspective and practical guidelines. *Industrial Management & Data Systems, 99*, 11–17.
Luna-Reyes, L. F., & Andersen, D. L. (2003). Collecting and analyzing qualitative data for system dynamics: Methods and models. *System Dynamics Review, 19*, 271–296.
Lynch, R. G. (2004). *Economic, fiscal, and social benefits of investment in early childhood development*. Economic Policy Institute.
Maharrani, A., & Syaifudin, N. (2020). *Ada asa pengelolaan sampah di Surabaya*. Retrieved September 4, 2021, from https://lokadata.id/artikel/ada-asa-kelola-sampah-di-surabaya
Mailena, L., Indrawanto, C., & Astuti, E. P. (2021). Risk management of chilli supply chains using weighted failure mode effect analysis. *IOP Conference Series: Earth and Environmental Science, 782*, 022004
Marobela, M. (2008). New public management and the corporatization of the public sector in peripheral capitalist countries. *International Journal of Social Economics, 35*(6), 423–434.
Matsui, Y., Nguyen, M. H., & Phan, A. C. (2018). Supply chain management in developing countries: Empirical evidence from Vietnamese manufacturing companies. *International Journal of Productivity and Quality Management, 24*(4), 566–582.
McKinsey. (2020). *How COVID-19 has pushed companies over the technology tipping point—And transformed business forever*. https://www.mckinsey.com/business-functions/strategy-and-corporate-finance/our-insights/how-covid-19-has-pushed-companies-over-the-technology-tipping-point-and-transformed-business-forever
Mcmichael, A. J., Butler, C. D., & Folke, C. (2003). New visions for addressing sustainability. *Science, 302*, 1919–1920.
Mendy Bilek, G., Fabbes-Coste, N., Kacioui-Maurin, E., Lazzeri, J., & Roussat, C. (2017). Obligation d'affichage des informations CO_2 transport: de la règlementation aux pratiques? *Logistique & Management, 2*.
Meyer, J. W., & Rowan, B. (1977). Institutional organizations: Formal structure as myth and ceremony. *American Journal of Sociology, 83*(2), 340–363.
Mihai, F. C. (2020). Assessment of COVID-19 waste flows during the emergency state in Romania and related public health and environmental concerns. *International Journal of Environmental Research and Public Health, 17*(15), 5439.
Min, S., Zacharia, Z. G., & Smith, C. D. (2019). Defining supply chain management: In the past, present and future. *Journal of Business Logistics, 40*(1), 44–55.

Mokhtarzadeh, N. G., Mahdiraji, H. A., Jafari-Sadeghi, V., Soltani, A., & Kamardi, A. A. (2020a). A product-technology portfolio alignment approach for food industry: A multi-criteria decision making with z-numbers. *British Food Journal, 122*, 3947–3967.

Mokhtarzadeh, N. G., Mahdiraji, H. A., Jafarpanah, I., Jafari-Sadeghi, V., & Cardinali, S. (2020b). Investigating the impact of networking capability on firm innovation performance: Using the resource-action-performance framework. *Journal of Intellectual Capital, 21*, 1009–1034.

Moosivand, A., Ghatari, A. R., & Rasekh, H. R. (2019). Supply chain challenges in pharmaceutical manufacturing companies: Using qualitative system dynamics methodology. *Iranian Journal of Pharmaceutical Research: IJPR, 18*(2), 1103.

Mulyanti, K., & Fachrurozi, A. (2016). Analisis sikap dan perilaku masyarakat terhadap pelaksanaan program bank sampah [studi kasus masyarakat Kelurahan Bahagia Bekasi Utara]. *Jurnal Ilmiah Ekonomi Manajemen dan Kewirausahaan "Optimal", 10*(2), 185–198.

Munir, M., Jajja, M. S. S., Chatha, K. A., & Farooq, S. (2020). Supply chain risk management and operational performance: The enabling role of supply chain integration. *International Journal of Production Economics, 227*, 107667.

Munro, D. A., & Holdgate, M. W. (1991). *Caring for the earth: A strategy for sustainable living*. International Union for the Conservation of Nature and Natural Resources.

Mustamu, R. H. (2007). Manajemen rantai pasokan industri farmasi di Indonesia. *Jurnal Manajemen dan Kewirausahaan (or) Journal of Management and Entrepreneurship, 9*(2), 99.

Narayana, S. A., Pati, R. K., & Vrat, P. (2014). Managerial research on the pharmaceutical supply chain—A critical review and some insights for future directions. *Journal of Purchasing and Supply Management, 20*(1), 18–40.

Nassereddine, H., Seo, K. W., Rybkowski, Z. K., Schranz, C., & Urban, H. (2021). Propositions for a resilient, post-COVID-19 future for the AEC industry. *Frontiers in Built Environment, 7*, 687021.

Niles, M. T., Bertmann, F., Belarmino, E. H., Wentworth, T., Biehl, E., & Neff, R. (2020). The early food insecurity impacts of covid-19. *Nutrients, 12*.

Norman, W., & Macdonald, C. (2004). Getting to the bottom of "triple bottom line". *Business Ethics Quarterly*, 243–262.

Novitasari, M., & Agustia, D. (2021). Green supply chain management and firm performance: The mediating effect of green innovation. *Journal of Industrial Engineering and Management, 14*(2), 391–403.

Obande, G. A., Bagudo, A. I., Mohamad, S., Deris, Z. Z., Harun, A., Yean, C. Y., Aziah, I., & Singh, K. K. B. (2021). The current state of the covid-19 pandemic in Africa: Lessons for today and the future. *International Journal of Environmental Research and Public Health, 18*, 1–15.

OECD. (2018). *Excessive pricing in pharmaceutical markets—Note by Indonesia*. https://one.oecd.org/document/DAF/COMP/WD(2018)114/en/pdf

OECD. (2020). *Covid-19 in Africa: Regional socio-economic implications and policy priorities*. https://read.oecd-ilibrary.org/view/?ref=132_132745-u5pt1rdb5x&title=COVID-19-in-Africa-Regional-socio-economic-implications-and-policy-priorities&_ga=2.195473004.1388414703.1636511120-1376897899.1636511120

OECD/ITF. (2015). *ITF transport outlook 2015*. OECD Publishing.

Ozkan-Ozen, Y., Kazancoglu, Y., & Kumar Mangla, S. (2020). Synchronized barriers for circular supply chains in industry 3.5/industry 4.0 transition for sustainable resource management. *Resources, Conservation and Recycling, 161*, 104986.

Peacocka, C., & Shermanb, D. M. (2010). Sustainable goat production: Some global perspectives. *Small Ruminant Research*, 70–80.

Popfalushi, D., & Lviv, Z. W. (2021). *The impact of the COVID-19 pandemic on the waste management system in Ukraine and what problems in this area have been identified*. Retrieved September 4, 2021, from https://zerowasteeurope.eu/2021/04/the-impact-of-the-covid-19-pandemic-on-the-waste-management-system-in-ukraine/

Popli, K., Sudibyo, G. L., & Kim, S. (2017). A review of solid waste management using system dynamics modeling. *Journal of Environmental Science International*, 1185–1200.

Prevent Epidemics. *Update on COVID-19 in Africa.*

Purwanto, B. (2020). Learning from the coronavirus pandemic: Interdisciplinary history and strategic issues of historical research. *Indonesian Historical Studies, 4*(2), 100–112.

Qazi, A., Simsekler, M. C. E., & Akram, M. (2021). Efficacy of early warning systems in assessing country-level risk exposure to COVID-19. *Geomatics, Natural Hazards and Risk, 12*, 2352–2366.

Qudrat-ullah, H. (2012). On the validation of system dynamics type simulation model. *Telecommunication Systems, 51*(2–3), 159–166.

Qudrat-Ullah, H., & Seong, B. S. (2010). How to do structural validity of a system dynamics type simulation model: The case of an energy policy model. *Energy Policy, 38*, 2216–2224.

Quinn, F. J. (1997). "What's the buzz?". *Logistics Management, 36*, 43–47.

Quynh, D. V. X., & Huy, N. H. (2018). Supply chain management practices, competitive advantages and firm performance: A case of small and medium enterprises (SMEs) in Vietnam. *Journal of Modern Accounting and Auditing, 14*(3), 136–146.

Rahman, M. R. (2013, February 1). *RE: Operation of Nahar Agro Complex Limited.* Type to Shamsuddoha, M.

Rahman, M. R. (2013, September 7). *RE: Modern Poultry operation.* Type to Shamsuddoha, M.

Rahman, S. (2012, January 25). Poultry litter can ease energy crisis. *The Daily Star.*

Rahmandad, H. (2015). Connecting strategy and system dynamics: An example and lessons learned. *System Dynamics Review, 31*(3), 149–172.

Ralston, P. M., Richey, R. G., & Grawe, S. J. (2017). The past and future of supply chain collaboration: A literature synthesis and call for research. *The International Journal of Logistics Management, 28*(2).

Ramanathan, U., & Gunasekaran, A. (2014). Supply chain collaboration: Impact of success in long-term partnerships. *International Journal of Production Economics, 147*, 252–259.

Rapitsenyane, Y. (2019). A conceptual review of sustainable innovation: A driver for growing the manufacturing industry in Botswana. *Journal of Creativity and Business Innovation, 5*, 43–61.

Reid, A., Ronda-Perez, E., & Schenker, M. B. (2021). Migrants workers, essential work, and covid-19. *American Journal of Industrial Medicine, 64*(2), 73–77.

Rejeb, A., Rejeb, K., & Keogh, J. G. (2020). Covid-19 and the food chain? Impacts and future research trends. *Scientific Journal of Logistics, 16*(4), 475–485.

Repenning, N. P. (2000). A dynamic model of resource allocation in multi-project research and development systems. *System Dynamics Review, 16*(3), 173–212.

Resource Center. (2019). *What is household waste?* Retrieved October 7, 2021, from https://www.buschsystems.com/resource-center/knowledgeBase/glossary/what-is-household-waste

Reynolds, P., Bygrave, W., Autio, E., & Cox, L. (2001). M. Hay, 2002. *Global entrepreneurship monitor, 2002 executive report.*

Rhyne, E. (1998). The yin and yang of microfinance: Reaching the poor and sustainability. *MicroBanking Bulletin, 2*, 6–8.

Richardson, G. P., & Pugh III, A. L. (1981). *Introduction to system dynamics modeling with DYNAMO.* Productivity Press.

Richardson, G. P., & Pugh III, A. L. (1986). *Introduction to system dynamics modelling with dynamo.* The MIT Press.

Roberts, N. H., Andersen, D. F., Deal, R. M., Grant, M. S., & Shaffer, W. A. (1983). *Introduction to computer simulation: The system dynamics modeling approach.* Addison-Wesley.

Ruele, M. (2011). Eradicating poverty and promoting dignity in Botswana through contextual theology of liberation: Challenges and prospects. *Journal of Social Development in Africa, 26*(1), 161–186.

Ruutu, S., Casey, T., & Kotovirta, V. (2017). Development and competition of digital service platforms: A system dynamics approach. *Technological Forecasting and Social Change, 117*, 119–130.

Saaty, T. L. (1974). Measuring fuzziness of sets. *Journal of Cybernetics, 4*, 53–61.

Saaty, T. L. (1980). *The analytic hierarchy process.* McGraw-Hill.

Saaty, T. L. (1986). Absolute and relative measurement with the AHP: The most livable cities in the United States. *Socio-Economic Planning Sciences, 20*, 327–331.

Saaty, T. L. (2009). *Theory and applications of the analytic network process: Decision making with benefits, opportunities, costs, and risks*. RWS.

Saaty, T. L. (2013). *Mathematical principles of decision making*. RWS.

Saeed, K. A., Malhotra, M. K., & Abdinnour, S. (2019). How supply chain architecture and product architecture impact firm performance: An empirical examination. *Journal of Purchasing and Supply Management, 25*(1), 40–52.

Saleque, M. A. (2013, June 12). *RE: Bangladesh Poultry and its future*. Type to Shamsuddoha, M.

Salomon, V. A. P. (2016). Absolute measurement and ideal synthesis in the AHP. *International Journal of the Analytic Hierarchy Process, 3*, 538–545.

Salyer, S. J., Maeda, J., Sembuche, S., Kebede, Y., Tshagela, A., Moussif, M., Ihekweazu, C., Mayet, N., Abate, E., Ouma, A. O., & Nkengasong, J. (2021). The first and second waves of the covid-19 pandemic Africa: A cross-sectional study. *The Lancet, 397*(10281), 1265–1275.

Sancheta, L. d., Chaves, G. d., & Siman, R. R. (2021). The use of system dynamics on urban solid waste management: A literature analysis. *Gestão & Produção*, 1–18.

Sánchez-Flores, R. B., Cruz-Sotelo, S. E., Ojeda-Benitez, S., & Ramírez-Barreto, M. E. (2020). Sustainable supply chain management—A literature review on emerging economies. *Sustainability, 12*(6972), 1–27.

Sargent, R. G. Verification and validation of simulation models. In *Proceedings of the 37th conference on winter simulation, 2005. Winter simulation conference*, 130–143.

Sarkis, J., Cohen, M. J., Dewick, P., & Schröder, P. (2020). A brave new world: Lessons from the COVID-19 pandemic for transitioning to sustainable supply and production. *Resources, Conservation & Recycling, 28*, 104894.

Sarkis, J., Gonzalez-Torre, P., & Diaz, B. (2010, March). Stakeholder pressure and the adoption of environmental practices: The mediating effect of training. *Journal of Operations Management, 28*(2), 163–176.

Sarkodie, S. A., & Owusu, P. A. (2020). Impact of COVID-19 pandemic on waste management. *Environment, Development and Sustainability*, 1–10.

Savitz, A. W., & Weber, K. (2006). *The triple bottom line*. Jossey-Bass.

Scholtens, B., & Kleinsmann, R. (2011). Incentives for subcontractors to adopt CO_2 emissions reporting and reduction techniques. *Energy Policy, 39*(3), 1877–1883.

Scuotto, V., Caputo, F., Villasalero, M., Del Giudice, M. (2017). A multiple buyer-supplier relationships in the context of SMEs' digital supply chain management. *Production Planning & Control, 28*, 1378–1388.

Sebata, D. (2021). *Supply chain management best practices: An empirical research on the Botswana state-owned enterprises* (Unpublished doctoral dissertation). Monarch Business School, Switzerland.

Seelos, C., & Mair, J. (2005). Social entrepreneurship: Creating new business models to serve the poor. *Business Horizons, 48*, 247–252.

Seelos, C., & Mair, J. (2007). Profitable business models and market creation in the context of deep poverty: A strategic view. *The Academy of Management Perspectives, 21*, 49–63.

Setino, R., & Amba, I. M. (2016). Supply chain management practices in state-owned enterprises environment. *Risk Governance & Control: Financial Markets & Institutions, 6*(4), 380–391.

Shamsuddoha, M., Quaddus, M. A., & Woodside, A. G. (2021). Environmental sustainability through designing reverse logistical loops: Case research of poultry supply chains using system dynamics. *Journal of Business & Industrial Marketing*.

Sheu, C., Yen, H. R., & Chae, B. (2006). Determinants of supplier-retailer collaboration: Evidence from an international study. *International Journal of Operations & Production Management, 26*(1), 24–49.

Shih, W. C. (2020). Global supply chains in a post-pandemic world. *Harvard Business Review, 98*(5), 82–89.

Sikdar, S. K. (2003). Sustainable development and sustainability metrics. *AIChE Journal, 49*, 1928.

Simão, L. E., Gonclaves, M. B., Rodriguex, C. M. T. (2016). An approach to assess logistics and ecological supply chain performance using postponement strategies. *Ecological Indicators, 63*, 398–408.

Sodhi, M. S., Son, B. G., & Tang, C. S. (2012). Researchers' perspectives on supply chain risk management. *Production and Operations Management, 21*(1), 1–13.

Sodhi, M. S., Tang, C. S., & Willenson, E. T. (2021). Research opportunities in preparing supply chains of essential goods for future pandemics. *International Journal of Production Research*. https://doi.org/10.1080/00207543.2021.1884310

Soosay, C. A., & Hyland, P. (2015). A decade of supply chain collaboration and directions for future research. *Supply Chain Management: An International Journal, 20*(6), 613–630.

Srivastava, R. K., Shervani, T. A., & Fahey, L. (1999). Marketing, business processes and shareholder value: An organizationally embedded view of marketing activities and the discipline of marketing. *Journal of Marketing, 63*(Special issue), 168–179.

Sterman, J. (2000). *Business dynamics: Systems thinking and modeling for a complex world*. Irwin/McGraw-Hill.

Sterman, J. D. (1989). Modeling managerial behavior: Misperceptions of feedback in a dynamic decision making experiment. *Management Science, 35*(3), 321–339.

Sterman, J. D. (2000). *Business dynamics: System thinking and modeling for a complex world*. Jeffrey J. Shelstad.

Sterman, J. D. (2000). *Business dynamics: System thinking and modeling for a complex world*. Jeffrey J. Shelstad.

Sterman, J. D. (2000). *Business dynamics: Systems thinking and modeling for a complex world*. Irwin McGraw-Hill.

Sterman, J. D. (2001). System dynamics modeling. *California Management Review, 43*, 8–25.

Steve. (2020). *7 Different types of rubbish you need to dispose of*. Retrieved October 11, 2021, from https://www.stevesrubbishremovals.com.au/7-different-types-of-rubbish-you-need-to-dispose-of/

Stock, J. H., & Watson, M. W. (2003). Has the business cycle changed and why? In *NBER Macroeconomics Annual 2002* (Vol. 17). MIT Press.

Sundin, E. (2011). Entrepreneurship and social and community care. *Journal of Enterprising Communities: People and Places in the Global Economy, 5*, 212–222.

Surabaya City Government. (2021). *Sebulan bisa capai 863 kilogram, begini cara pemkot tangani sampah rumah tangga masker*. Retrieved September 4, 2021, from https://www.surabaya.go.id/id/berita/61997/sebulan-bisa-capai-863-kilogram

Svensson, G. (2007). Aspects of sustainable supply chain management (SSCM): Conceptual framework and empirical example. *Supply Chain Management, 12*, 262–266.

Tan, K. C., & Blanco, E. E. (2009). *System dynamics modeling of the smartway transport partnership*. MIT Engineering Systems Division & CESUN. Presented at the Second International Symposium on Engineering Systems at the MIT Center for Transportation and Logistics, Cambridge, Massachusetts (June).

Tarifa-Fernandez, J., De Burgos-Jiménez, J. (2017). Supply chain integration and performance relationship: A moderating effects review. *The International Journal of Logistics Management*.

Tasrif, M. (2004). *Model simulasi untuk analisis kebijakan: pendekatan metodologi system dynamics*. Institut Teknologi Bandung.

Tate, W. L., Ellram, L. M., & Dooley, K. J. (2012). Environmental purchasing and supplier management (EPSM): Theory and practice. *Journal of Purchasing and Supply Management, 18*(3), 173–188.

Täuscher, K., & Laudien, S. M. (2018). Understanding platform business models: A mixed-methods study of marketplaces. *European Management Journal, 36*, 319–329.

Temtime, Z. T., Chinyoka, S., & Shunda, J. (2004). A decision tree approach for integrating small business assistance schemes. *Journal of Management Development, 23*(6), 563–578.

The Economist. (2021). *How covid-19 is boosting innovation?* https://www.economist.com/films/2021/03/10/how-covid-19-is-boosting-innovation

Tilman, D., Cassman, K. G., Matson, P. A., Naylor, R., & Polasky, S. (2002). Agricultural sustainability and intensive production practices. *Nature, 418*, 671–677.

Tilman, D., Wedin, D., & Knops, J. (1996). *Productivity and sustainability influenced by biodiversity in grassland ecosystems*.

United Nations. (2020). *Policy brief: The world of work and covid-19*, 1–24. United Nations. https://www.un.org/sites/un2.un.org/files/the_world_of_work_and_covid-19.pdf

Van Hoek, R., Gibson, B., & Johnson, M. (2020). Talent management for a post-covid-19 supply chain—The critical role for managers. *Journal of Business Logistics, 41*(4), 334–336.

Velayutham, A., Rahman, A. R., Narayan, A., & Wang, M. (2021). Pandemic turned into pandemonium: The effect on supply chains and the role of accounting information. *Accounting, Auditing & Accountability Journal, 34*, 1404–1405.

Venkataraman, S. (2002). *The distinctive domain of entrepreneurship research*. Edward Elgar Press.

Vennix, J. A. M. (1996). Group model building: Facilitating team learning using system dynamics. Wiley.

Vieira, A. J. (2020). Supply chain disruptions and challenges post covid 19 crises in the Indian context. *Supply Chain Pulse, 11*(1), 22–23.

Wagner, J. (2003). Testing Lazear's jack-of-all-trades view of entrepreneurship with German micro data. *Applied Economics Letters, 10*, 687–689.

Wakeford, J. J. (2012). *Socioeconomic implications of global oil depletion for South Africa: Vulnerabilities, impacts and transition to sustainability*. Stellenbosch University.

Walters, J. P., Archer, D. W., Sassenrath, G. F., Hendrickson, J. R., Hanson, J. D., Halloran, J. M., … Alarcon, V. J. (2016). Exploring agricultural production systems and their fundamental components with system dynamics modelling. *Ecological Modelling, 333*, 51–65.

Wan, C., Shen, G. Q., & Choi, S. (2019). Waste management strategies for sustainable development. In W. Leal Filho (Ed.), *Encyclopedia of Sustainability in Higher Education*, 1–9.

Wang, M., & Jie, F. (2019). Managing supply chain uncertainty and risk in the pharmaceutical industry. *Health Services Management Research, 33*(3), 156–164.

Wang, M., Wang, B., Song, B., & Bayne, K. (2020, June). Understanding effects of COVID-19 in New Zealand businesses: An early-stage study. Scion*: Business Outlook*, 1–8.

Wang, W., Tang, J., & Wei, F. (2020). Updated understanding of the outbreak of 2019 novel coronavirus (2019-nCov) in Wuhan, China. *Journal of Medical Virology, 92*, 441–447.

Ward, L. (2002). *Environmental policies for a sustainable poultry industry in Sussex county, Delaware*. University of Delaware.

4waste. (2016). *There are 5 types of waste, do you know them all?* Retrieved October 10, 2021, from https://4waste.com.au/rubbish-removal/5-types-waste-know/

WCED. (1987). *Our Common Future* (The Brundtland Report). Oxford University Press.

WHO. (2020). *Coronavirus disease (COVID-19) pandemic*. https://www.who.int/emergencies/diseases/novel-coronavirus-2019

WHO. (2020). *Water, sanitation, hygiene, and waste management for SARS-CoV-2, the virus that causes COVID-19*. World Health Organization and the United Nations Children's Fund (UNICEF).

Wolmarans, P., Hyland, E., Atherton, S., Bovet, D., Bryan J., & Cheng, A. (2014). *Sustainability strategies addressing supply chain air emissions*. National Cooperative Freight Research Program, 28.

Wolmarans, P., Hyland, E., Atherton, S., Bovet, D., Bryan, J., & Cheng, A., (2014). *Sustainability strategies addressing supply-chain air emissions*. National Cooperative Freight Research Program and the Transportation Research Board of National Academies, 28.

World Bank. (2020). The global economic outlook during the COVID-19 pandemic: A changing world. https://www.worldbank.org/en/news/feature/2020/06/08/the-global-economic-outlook-during-the-covid-19-pandemic-a-changed-world

World Bank. (2021a). World development indicators. http://wdi.worldbank.org/

World Bank. (2021b). Pandemic preparedness and COVID-19 (coronavirus). https://www.worldbank.org/en/topic/pandemics

Yanamandra, R. (2021). Investigating the influence of organizational factors on supply chain awareness. *Operations and Supply Chain Management, 14*(2), 189–202.

Yu, K. D. S., & Aviso, K. B. (2020). Modelling the economic impact and ripple effects of disease outbreaks. *Process Integration and Optimization for Sustainability, 4*, 183–186.

Yu, X., Li, C., Shi, Y., Yu, M. (2010). Pharmaceutical supply chain in China: Current issues and implications for health system reform. *Health Policy, 97*(1), 8–15.

Yunus, E. N., & Kurniawan, T. (2015). Revealing unsuccessful collaboration: A case of buyer-supplier relationship in the pharmaceutical industry. *Supply Chain Management: An International Journal, 16*(2), 14–28.

Yunus, E. N., & Tadisina, S. K. (2016). Drivers of supply chain integration and the role of organizational culture: Empirical evidence from Indonesia. *Business Process Management Journal, 22*(1), 89–115.

Yunus, M. (2007). *Creating a world without poverty: Social business and the future of capitalism.* PublicAffairs Store.

Zebda, A. (2002, September). Using cost-benefit analysis for evaluating decision models in operational research. *Journal of American Academy of Business*, 106–114.

Zekhnini, K., Cherrafi, A., Bouhaddou, I., Benghabrit, Y., & Garza-Reyes, J. A. (2020). Supply chain management 4.0: A literature review and research framework. *Benchmarking: An International Journal, 28*(2), 465–501.

Zhu, Q., & Sarkis, J. (2007). The moderating effects of institutional pressures on emergent green supply chain practices and performance. *International Journal of Production Research, 25*(18–19), 4333–4355.

Zidi, S., Hamani, N., & Kermad, L. (2021). New metrics for measuring supply chain reconfigurability. *Journal of Intelligent Manufacturing*. https://doi.org/10.1007/s10845-021-01798-9

Zoumpourlis, V., Goulielmaki, M., Rizos, E., Baliou, S., & Spandidos, D. A. (2020). The covid-19 pandemic as a scientific and social challenge in the 21st century. *Molecular Medicine Reports, 22*, 3035–3048.

Index

H. Qudrat-Ullah (ed.), *Understanding the Dynamics of New Normal for Supply Chains*, Understanding Complex Systems,
https://doi.org/10.1007/978-3-031-07333-5

CPSIA information can be obtained
at www.ICGtesting.com
Printed in the USA
LVHW082330111022
730433LV00016B/9

9 783031 073328